DIY

Lithium Batteries

The Ultimate Guide to Understanding Lithium Batteries and How to Make a Lithium Battery Pack for Electric Bikes

© Copyright by **Ryan Greene**

Table of Contents

INTRODUCTION

Lithium batteries are extremely hazardous. They hold a lot of energy in a tiny amount of space and are designed to release that energy quickly. They can be a safe and efficient way to power practically anything if utilized correctly. Lithium batteries, on the other hand, have the potential to produce disastrous fires, resulting in the loss of property and lives if used incorrectly.

This manual is meant to be instructional. Please get professional assistance and training before attempting to duplicate anything you see in this book. Always wear the proper safety gear. Never leave a lithium battery charging station unattended. Always have your wits about you, be clever, and stay safe.

CHAPTER 1 – Lithium Battery Cells

In the 1970s, Whittingham invented the first lithium battery. He found how to store energy in rechargeable batteries using pure lithium (the lightest metal in nature) in Exxon's labs. Unfortunately, due to lithium's instability, the battery is unworkable and could explode.

Goodenough improved Whittingham's idea in the 1980s, eventually creating Sony's first Li-Ion battery in 1991 by changing the cathode composition (from disulphide to cobalt oxide) and increasing the battery's power.

Akira Yoshino, on the other hand, eliminates lithium in its pure form from the battery and relies solely on lithium ions for functioning, making the batteries safer and more widely usable. Whittingham's prototype is turned into a commercial device by Goodenough and Yoshino, which increases power and safety.

Apart from scientists, he is the true Nobel Prize winner: lithium, a metal and the lightest solid element. Lithium in the form of an ion is more frequent than pure lithium because it is more stable (positively charged, due to the loss of an electron). Nobel batteries, which have a high energy density and can be recharged several times without losing their effectiveness, are built around this positively charged particle. The ability of lithium ions to flow across battery electrodes without chemical reactions modifying them has enabled a revolution in the storage and application of electrical energy (the essence of batteries). Because of their propensity to intercalate, they can be contained within the electrode materials. A conquest that took place, step by step.

The functioning of lithium batteries

Whittingham's battery functioned like this, as indicated in the diagram: lithium ions (along with electrons) were transported from the anode to the cathode via titanium disulphide, and then returned when the battery was recharged.

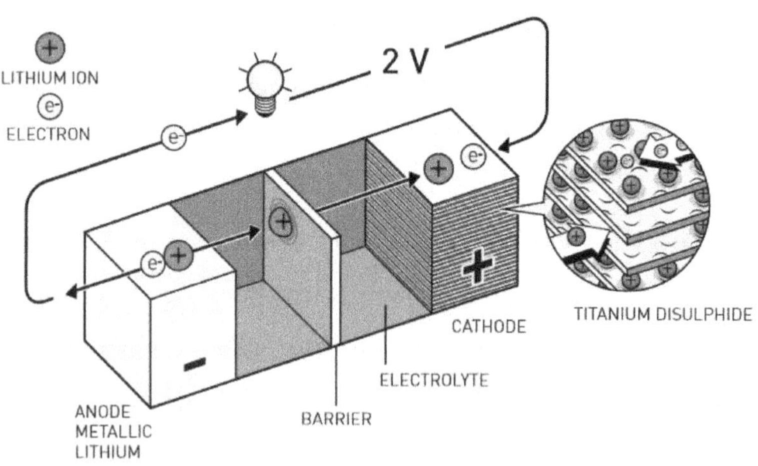

Whittingham's battery, however, lacked practicality: the lithium metal used in the anode was too explosive. Aluminium was added and the electrolyte was modified, but lower crude oil prices made Exxon's interest in field research obsolete, and so Whittingham's work was interrupted. Only with Goodenough's arrival on the scene would lithium batteries begin to regain interest and become what they are today. During the eighties Goodenough in fact sensed that changing the cathode constitution could increase the power of the batteries: by replacing titanium disulphide with cobalt oxide the Oxford researcher was able to develop a 4-volt battery, twice the size of Whittingham's one.

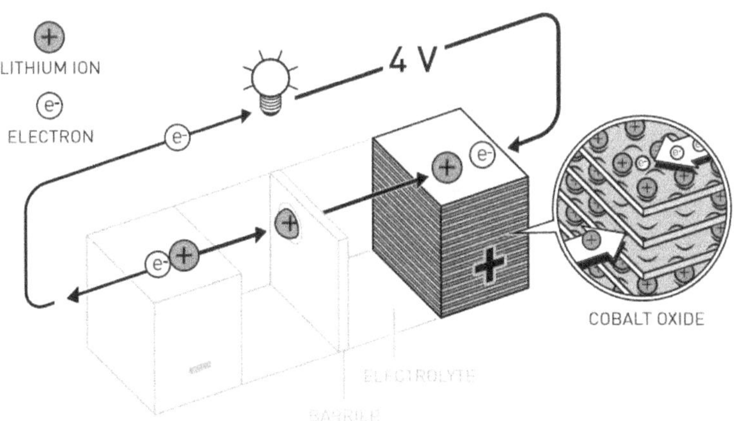

It would be the need for light and powerful batteries from the Far East that would give a decisive boost to research into lithium-ion batteries. Akira Yoshino, the third Nobel Prize winner in chemistry, thought of using petroleum coke (a by-product of product processing) for the construction of the anode material to house lithium ions, in a similar way to cobalt oxide in the cathode. The result was a stable, lightweight and safe product, ready since the mid-1980s to become a commercial product.

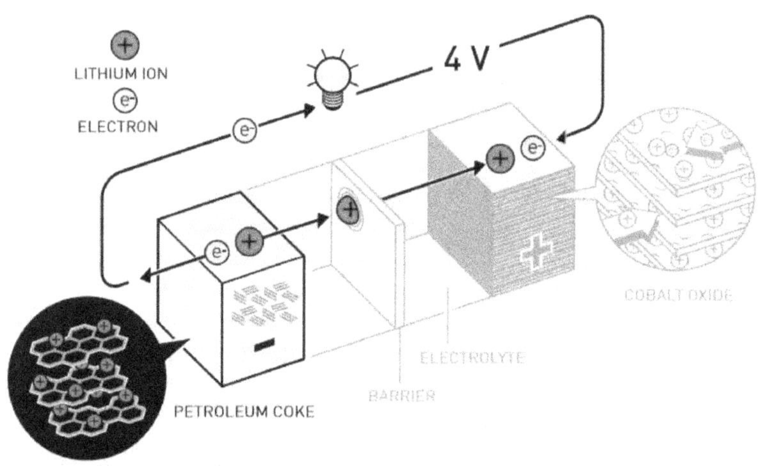

Lithium Batteries

The Lithium Ion cells, whose number depends on the device to which they are connected, are composed of a complex inner layer of Carbon and Lithium, interconnected by a layer of highly conductive material. The cells are pressurized inside a metal casing, through which the temperature increase or decrease can be monitored. The sensor, called PTC - Positive Temperature Coefficient placed above the cell, detects and allows you to stop using the cell when the internal temperature rises above a certain value.

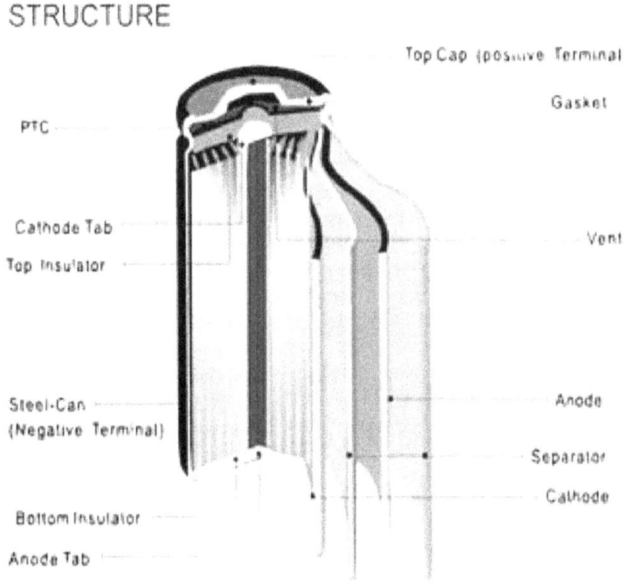

The Carbon and Lithium layers, on the other hand, represent the heart of the operation of Lithium batteries. They are the anode and cathode of the battery. Thanks to their chemical characteristics they have the ability to "attract" and "hold" the charge for a very long time. In this sense, we can assimilate their action to that of the plates of a common capacitor, but with much more interesting characteristics.

The Carbon (negative) and Lithium Cobalt Oxide (positive) layers are two electrodes that lend themselves very well to this type of situation, in relation to their chemical qualities. In fact, when immersed in an electric field (called radiation energy), they release Ions, electrons that detach from atoms and become free. When the battery is connected to a power source, in fact, the electrons of the Lithium layer leave the atoms and pass through the shooting layer (SEI), made of polymeric gel electrolyte material. The movement of these electrons is so fast as to produce a very high voltage especially in relation to the very small space in which the phenomenon occurs (nanometres, one billionth of a meter). The voltage inside the cell is estimated to be around 3.7 Volts. With this continuous voltage the cell produces enough energy to supply normal electronic circuits. Much higher voltages can be achieved through parallel systems.

Once the carbon layer is reached, the electrons "connect" to its atoms, remaining blocked due to electrical forces generated internally. The charging process ends when the "ionization", the term used to identify the ion transfer process, is complete. It is important to stress that it is the radiation energy that allows the ions to move in a precise direction, from the anode to the cathode.

During the discharge phase, when the device is in operation, in fact, the radiation energy, internal to it, determines the displacement in the opposite direction, from the cathode (Carbon) to the anode (Lithium). We know, however, that the discharge phase is much slower than the charging phase, so that the Li-ion battery can be charged in a short time and last longer.

The discharge duration of the Li-Ion battery installed on modern portable devices such as smartphones, tablets and e-books naturally depend on the conditions of use of the device. The more energy the device requires to work, the faster the discharge phase is carried out, the faster the ions are bound back to the Lithium layer atoms. However fast it may be, however, the discharge phase will always be slower than the charging phase, as the Carbon is more resistant to ionization.

Difference between batteries: LI-ION, LIPO, NICD and NIMH

Reading the technical characteristics of the batteries, whether they are for smartphones, PCs, tablets, power tools or classic batteries, we will notice that in addition to codes, mAh and voltage there is also an abbreviation that indicates the type of batteries in use. LiPo, NiCd, NiMh and Li-Ion, are the types of batteries most used today. These abbreviations are nothing more than an indication of the materials from which the batteries are made.

Li-Ion, a powerful battery without memory

This particular type of battery saw the light in 1991 and is still used today for mobile phones and laptops.

These are very light batteries, which have had great success also thanks to lower production costs compared to Ni-Cd batteries.

Li-Ion batteries have many advantages, the first of all is the absence of the so-called memory effect that distinguishes many other rechargeable batteries. To explain it in very simple terms, it is possible to recharge a Li-Ion battery even when it is not completely discharged, without compromising anything. On the contrary for batteries that suffer from memory effect, if they are always recharged when they have a residual 10%, that 10% of charge will become (for memory effect) 0%. So, the battery will be discharged even if it has a residual 10%.

Among the other advantages there is certainly the possibility to subject the battery to several charge/discharge cycles. In addition, unlike a NiMh battery that if not used loses even 50% of its charge over time, lithium-ion batteries can be used even after months, without this leading to a considerable loss of charge.

One of the negative points of Li-Ion batteries is their particular sensitivity to heat. If overheated it can buy explosions and in any case, overheating leads to faster deterioration over time.

The battery that is difficult to deteriorate, the LiPo

The LiPo battery, or lithium polymer battery, is the evolution of Li-Ion and it is more resistant and less expensive to produce, details that should not be underestimated.

You will notice that the recharge cycle is higher than Li-Ion batteries (about 20%) and also are batteries that have a longer autonomy. They are smaller in size, do not suffer from memory effect and need less time to recharge.

However, if this is not charged with the correct adapter, charging will be slow enough.

NiMh, a battery with a big charge

The NiMh battery is made with an anode alloy, whose characteristic is to absorb hydrogen, making it more durable in terms of recharging.

This is one of the batteries with a higher capacity than the others: if, for example, a NiMh battery can reach a hypothetical value of one hundred, the NiCd battery can accumulate an energy equal to one hundred and twenty.

However, it should be noted that this battery suffers the most from the self-discharge problem.

It also tends to suffer minor damage if it is not constantly recharged.

The battery with large cycles, the NiCd

This rechargeable battery consists of a combination of nickel and cadmium.

This feature allows you to have a battery that is more durable over time, you can charge and discharge it for a number of times more than the NiMh battery.

It should be noted that this model is also resistant even if this is not used and also the discharge speed of the same is less immediate than the battery previously analysed.

The Flammability of Lithium Batteries

One of the problems with Lithium batteries is their high flammability due to their predisposition to explosion. Although this represents a very high danger to the health of users, it is important to stress that the safety measures taken in current batteries are sufficient to limit possible disasters.

As you probably know, in order to trigger a combustion reaction, it is necessary to have all three elements of the so-called combustion triangle: a comburent, usually oxygen, a fuel, any substance capable of igniting, and a source of ignition, which may be the heat input.

Inside the Lithium battery, the Lithium and Carbon layer are generally exposed to an average temperature of 45 °C, both during charging and discharging. The temperature increase and decrease within certain margins are controlled by the internal microcontroller when it is active. However, when an external agent is applied, whether it is a simple overcharging of the circuit or an increase in the external temperature of the battery, the internal temperature rises above the set limits. However, it is proven that once the threshold of 75 °C is exceeded, the internal overheating reaction becomes so autonomous that it generates a non-return phenomenon defined as Domino Effect in which, although acting on the battery, it is impossible to avoid the explosion.

The Domino Effect in fact consists in the generation of very high heat chain mixed with the production of internal gases, which determine a pressure so strong that it breaks the outer coating, and generate the explosion. The separating electrolyte material then, containing lithium ions at high temperature, coming into contact with oxygen (comburent) generates the inevitable combustion reaction.

CHAPTER 2 – Lithium Battery Composition

Accumulators

Accumulators, or rechargeable batteries, or secondary batteries (usually called batteries), are electrochemical devices capable of accumulating electrical charge and returning it during the discharge phase.

They consist of a set of elementary cells, electrically connected to each other in series or in parallel, to reach the desired value of voltage (Volt) and current (Ampere).

The values that characterize a battery - for example the starting one of a car - are:

Rated voltage: it is expressed in Volts and indicates the work that can be done by the electric charging unit when switching from positive to negative pole. It is an indicator of the degree of danger inherent in electrical energy: in fact, voltages above 60 V are dangerous for the human body. This is why special care is required when using electrical equipment.

Nominal capacity: it is expressed in Ah and represents, in a simplified way, the number of amperes that the battery can deliver for one hour in a row.

Maximum discharge: it is expressed in Amperes and is the maximum current that the battery can deliver for a few seconds (for example for starting the car engine).

Technology used: it is indicated with abbreviations such as Pb, Li, etc.

Geometric measurements: these are important when the housing, as in the car, has a mandatory size.

When a single battery has insufficient characteristics for the required use, it is necessary to assemble several batteries. The use of several batteries at the same time is done in series or parallel assembly.

Parallel connection   **Series connection**

By properly connecting the elements, battery packs can be built to suit a wide range of applications, from low power and low capacity for medical or consumer electronics to high power and high capacity for electric and hybrid vehicles.

Even larger batteries can be connected in series and parallel.

Connecting the elements in series (plus/minus) increases the voltage. The total voltage is the sum of the voltages of each element. In the example in the figure, the total voltage of two 12 V batteries connected in series is 24 V.

SERIAL CONNECTION

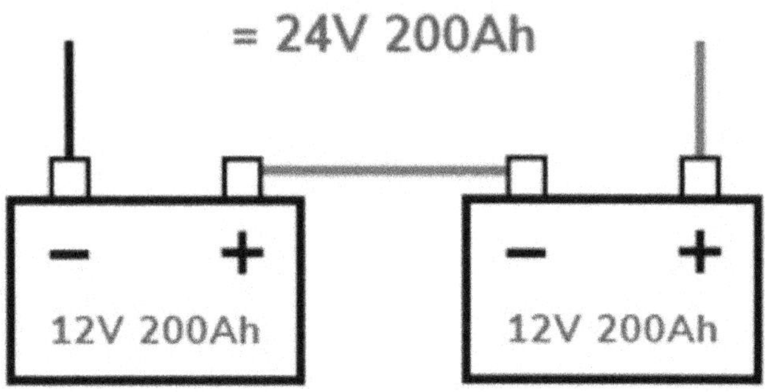

Connecting the elements in parallel (more with more, less with less) increases capacity. The total capacity is the sum of the capacities of each element. In the example in the figure above, the total capacity of two 200 Ah batteries connected in parallel is 400 Ah.

PARALLEL CONNECTION

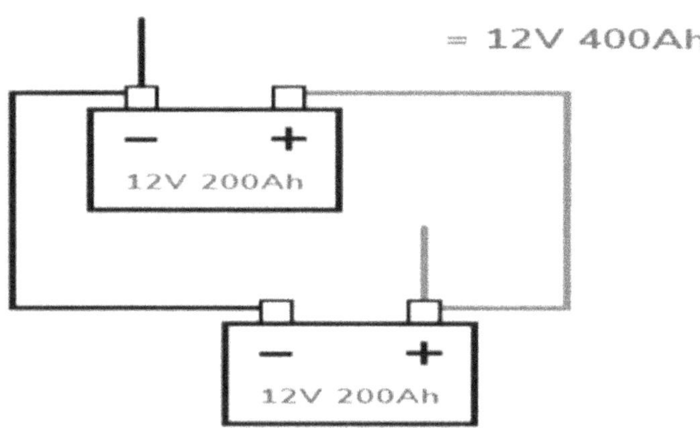

By combining the two methods you can do all combinations of voltage and capacitance.

The elements in the figure below are horizontally connected in series. The two resulting packages have been connected (vertically) in parallel.

SERIES-PARALLEL CONNECTION

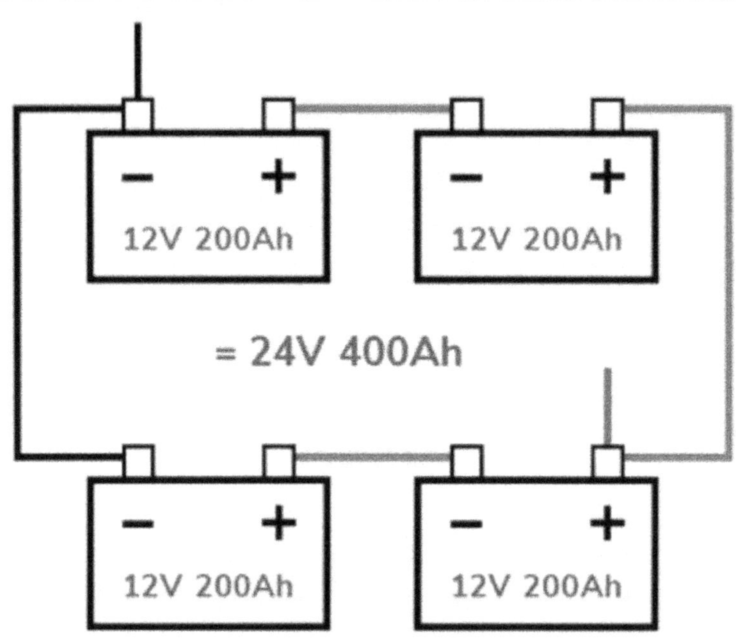

Compose a Battery Pack

A lithium battery pack consists of various components: the elementary cells, a container, an electronic supervision and control system called "Battery Management System" (BMS), fuses, terminals and connection cables, a cooling system, a data communication system.

Elementary cell: Lithium-ion cells are generally used for automotive batteries. There are other types of cells always based on lithium.

Connecting cables: they are made of copper; they can be flexible cables or rigid plates. They are used to connect all cells.

Signal cables: they are used to carry voltage, current and temperature measurements from each cell to the BMS.

Main and secondary modules of the BMS: this is the electronics of the package. It carries out checks for safety, for knowledge of residual charge, for performance optimization.

Traction current cables: large copper cables. They carry current from the batteries to the electric drive (inverter and motor).

Vehicle communication interface: information from the battery is fed into the vehicle's CAN communication system.

Typical information sent are: amount of charge available, expected autonomy, temperature, various alarms.

Current meter: instrument for measuring and controlling the current supply.

Insulation meter: instrument that verifies electrical safety.

Main relay: it is an automatic switch for manoeuvring and safety. It separates the batteries from the rest of the car.

What is an elementary cell?

The cell is the elementary construction element of the batteries. It is the minimum device that converts the chemical energy, contained in active materials, directly into electrical energy by means of electrochemical oxidation-reduction reactions.

A cell comprises two plates - electrodes - which are positively and negatively charged, immersed in a particular liquid - electrolyte or even electrolyte. Different reactions occur on each of the two plates, both during charging and discharge. During charging, the reactions convert the electric energy supplied from outside into chemical potential. The opposite is the case during discharge.

Rated voltage: The cell can be considered as an electron pump. The nominal voltage, or cell potential, is the voltage between the poles and corresponds to the "pressure" of the hypothetical electron pump.

Cell Components

The anode: during discharge the anode, negative pole, supplies electrons to the external circuit. Lithium oxidizes. It is generally made up mostly of graphite.

The cathode: during discharge the cathode, positive pole, accepts electrons from the external circuit. Lithium is subject to the reduction reaction. It consists of lithium oxide. The entire cell type is named after the material used for the cathode.

The electrolyte: is a conductor of ions (charged particle), but an insulator for electrons. It separates the two electrodes and provides the medium for the transfer of ions between the anode and cathode. In lithium batteries the electrolyte is an inorganic solvent containing lithium salts. A porous septum electrically separates the two electrodes but allows the ions to transit from one electrode to the other.

Charging phase: during charging the positive electrode is oxidized. The Li+ ions are subtracted from the lithium oxide, pass through the electrolyte and membrane and are placed between the graphite layers of the negated electrode with the negated electrode reduction reaction.

Discharge phase: An oxidation of the negative electrode takes place during the discharge. The Li+ ions are ripped from this and migrate through the electrolyte to install themselves in the positive electrode. An equivalent number of electrons travel through the external circuit producing the useful current (by convention the current is understood as a flow of positive charges, so the reverse is opposite to that of the negatively charged electrons).

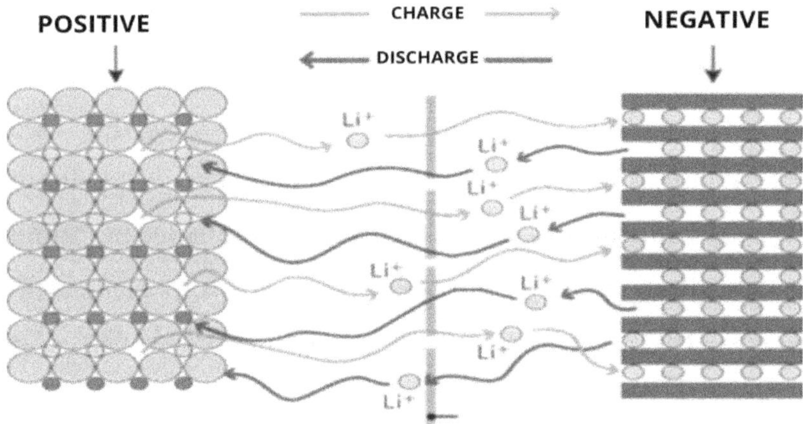

CHAPTER 3 - The Cells, one Shape for Every Use

An important element is the cell container; it is used to keep the components together, but from its choice follows a clear characterization of the cell.

Prismatic: a prismatic container made of aluminium or thin steel is typical of Li-ion type cells. The construction allows high energy density, excellent heat dissipation, excellent packaging and exploitation of space. They come up to the size of 100 Ah.

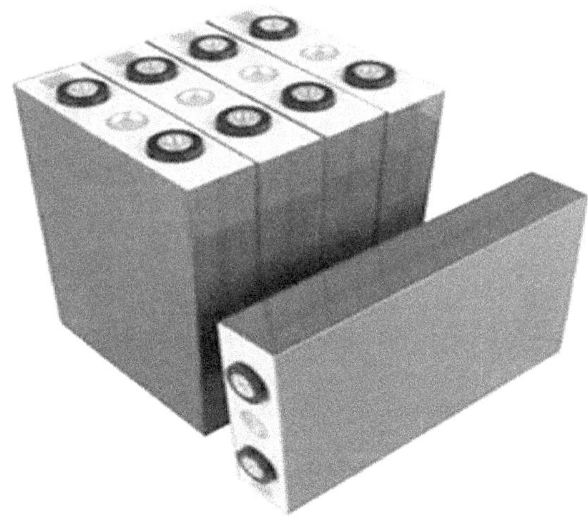

Advantages: The prismatic cell case is very resistant; they lend themselves very much to the creation of custom battery packs being able to avoid elaborate studies and expensive tests on mechanical stress.

High capacity of individual elements, you can get with a single cell up to 300Ah, this is very appreciated in the creation of batteries with important capacities such as the industrial market, avoiding parallel cells greatly increases the safety of the entire package.

Another important aspect are the screw poles, this simplifies assembly and makes it easy to replace an element, even on site.

Disadvantages: The disadvantage of prismatic lithium cells is the slightly lower energy density due to the type of housing.

Cylindrical: Small cylindrical cells are very common, for example, in packs for laptop batteries. There are up to 200 Ah on the market. Larger cells are not highly marketable and are therefore expensive.

Advantages: The advantages of using cylindrical lithium cells are the resistant case and the possibility to change chemistry while keeping the containment mechanics unchanged.

Disadvantages: The negative aspects of cylindrical lithium cells are the low capacity of the single cell, up to a maximum of 3Ah and the cells are assembled in series and in parallel by welding.

This type of construction does not allow the replacement of a single cell but requires the replacement of an entire module with more invasive and expensive interventions. If your industrial vehicle signals a problem on the battery, then you will have to replace the entire module and you will encounter costly downtime.

In addition to this, the low capacity of the individual elements and the resulting "parallel" of many elements often causes a drop in the safety of the entire lithium battery pack.

Example

Most lithium battery assemblers put N cells in parallel to achieve the desired capacity, for example if a capacity of 200 Ah is required and 2.5 Ah cells are used, they will have to put 80 cells in parallel, having all these cells connected in parallel. If one of these for an internal problem short-circuits it will not only absorb its energy but will also have to dissipate all the energy of the 80 cylindrical cells it has in parallel.

This phenomenon could turn into a very high module-side heat causing also disastrous effects like fire.

To avoid this, large manufacturers like Tesla use very sophisticated production processes that allow to connect the cells in parallel using a kind of fuse. If a cell goes short the "fuse" blows and the battery pack remains safe.

Unfortunately, however, there are no manufacturers of lithium batteries in the industrial market that have these technologies.

Envelope: thin envelope cells make the best use of space. They do not have a rigid container, so the weight is also very low. They can be manufactured in various shapes, even to order. They use the Lithium polymer technology; they are becoming very popular as an alternative to prismatic ones, especially for electric cars. The electrolyte is a polymer, so there is no leakage of liquids. Cooling is easy because of the large surfaces. However, the low mechanical resistance, which requires a suitable packaging system, should be highlighted.

Advantages: The positive aspect of this lithium technology is the high energy density that can be obtained and the low cost of the casing. You can find on the market pouch cells from a few hundred mAh up to 20Ah, there are very few manufacturers that exceed this threshold.

Disadvantages: The most negative aspect is certainly the delicacy of this casing, it takes very little to ruin it, even with a fingernail can be irreparably damaged.

In the use of this type is very important the packing system of the cells, which must prevent any kind of stress to the individual cells: vibrations, crushing and deformations.

The packing system of the cells for the creation of the modules is very important so it is complicated, if not impossible, to create custom modules.

The cells are always of low medium capacity.

To make industrial battery packs would need dozens of cells in parallel going to lower the safety of the battery pack. Also, in this case the individual cells are soldered to connect in series or parallel, and it is therefore impossible to replace a cell, you must replace the entire module.

Cell Chemistry

There are different ways to exploit the chemical reactions of the electrolytic cell; each of these corresponds to a different performance. Lithium cells have a higher energy density than previous technologies. They are safe, non-toxic, long-lived. They are the most suitable for electric and hybrid vehicles.

• Lead-acid (Pb-acid): lead acid cells are the oldest. They consist of a lead oxide cathode and a spongy lead anode. The electrolyte is a sulphuric acid solution diluted in distilled water.

The cell voltage is 2 V.

It is heavy, due to the very nature of the electrodes, and dangerous, due to the possible leakage of acid.

It has low specific energy, medium performance for power and low temperature operation, reasonably low self-discharge, difficult to monitor charge status, slow charging.

Metal components are recyclable but have low residual value; the presence of acid makes recovery difficult.

They are subject to material aging failure after less than 1000 charge and discharge cycles.

The lead-acid battery is widespread and produced on a large scale at low prices. Its main use is as a starter battery and for the on-board installation of vehicles and vessels. It is also used for emergency systems, as storage for isolated users, for uninterruptible power supply systems (anti blackout), for electric vehicles.

• Nickel-Cadmium (Ni-Cd): these cells use nickel oxide per cathode, cadmium per anode, salt per electrolyte. They have good characteristics, but they suffer from the phenomenon of "memory", so if you stop the discharge at a partial level several times, it will be stored as the last stage of discharge. These batteries were used in the eighties in the first portable video and telephony devices; they were soon banned because of the high toxicity of cadmium.

• Nickel-Metallic Hydrides (Ni-MH): these are the batteries that replaced the Ni-Cd batteries, and they are still very popular, especially in the stylus format. They use a nickel oxide for the cathode; the anode is a metal (lanthanum or other rare earth), with hydrogen adsorbed to the surface.

They have good charging capacity and speed and do not suffer from memory effect. For applications where cost is not an issue, they have been supplanted by lithium batteries.

• Sodium Nickel Chloride (Na-NiCl2): they are the evolution of sodium-sulphur technology; they use a salt always kept in a molten state at about 300 degrees, even if no dangerous temperature is felt outside the container. For its technology it can only be quite large, and in fact the only use is for the traction of electric cars. It must always remain at a constant temperature, so it dissipates some of its energy content for this purpose. It cannot be left without charging for a long time.

Lithium is attractive because of its light weight and electrical potential. The characteristics are: high charging capacity, high power, relatively low raw material cost, high charging speed, high number of cycles (life), high availability of raw material and low environmental impact for production and disposal. Various technologies are already available, and others are being studied.

Lithium Batteries

Lithium-ion Batteries

Lithium-ion technology uses a graphite anode and various components for the cathode. The electrolyte is a lithium salt dissolved in an anhydrous organic solvent. Current studies focus on the cathode structure.

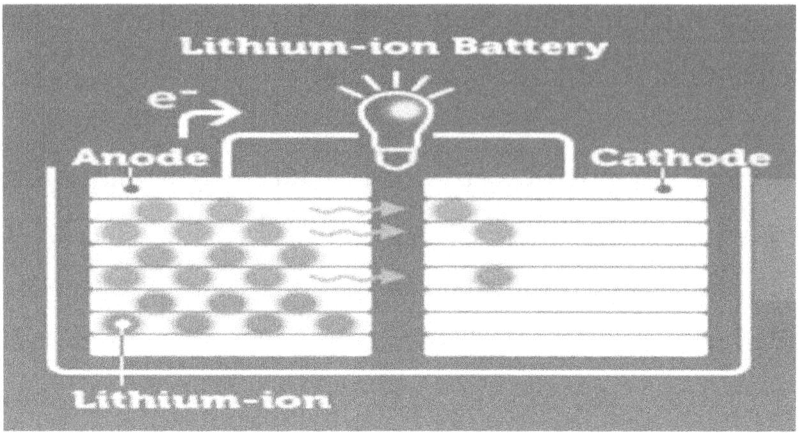

The technology of the first models included the lithium metal cathode. This was subject to an implicit chemical instability, during the recharge phase could incur overheating that led to lithium fusion and a fast and uncontrollable oxidation. In the years following 1991 a considerable amount of lithium batteries were withdrawn following the release of gas or the fire of some specimens. It is estimated that the specimens in failure were one in 200 thousand, and this forced the withdrawal of about six million packs, only by Dell and Apple. As a result of these failures, battery production focused on replacing the lithium metal cathode with a non-metallic lithium cathode, particularly the lithium ion cathode dispersed in a crystalline matrix of a stable oxide.

• Lithium - Cobalt (LiCoo2): is the most common type for cathode construction. Obviously, the cobalt used is the nonradioactive isotope. It is still used for laptops and mobile phones. It still suffers from heating problems, as well as some instability when the container is perforated. For these reasons it is the least used for electric traction.

- Lithium-iron-phosphate (LiFePo4): has superior thermal stability. Lithium phosphate is non-combustible and will not decompose under short circuit. It allows a high cyclic life (2000-3000 cycles). These batteries have lower energy than Lithium-Cobalt batteries, but higher power. They are still superior in safety, cost and toxicity.

- Lithium-Manganese Dioxide (LiMn2or4): offers high cell voltage and thermal stability, but slightly less energy than the others. The cost is low and the materials are not toxic. Good performance at high temperature.

- Lithium-Nickel-Cobalt Manganese LiNixCoyMnzo2 - NCM): they represent a good compromise between the various characteristics of the technologies

- Lithium Titanium Dioxide (Li4Ti5or12 - LTo): these cells replace the graphite anode with one of lithium titanate. This is typically used in conjunction with a manganese cathode. They offer satisfactory electrical characteristics and avoid the danger of graphite burning.

Lithium Polymer Batteries

The rechargeable battery known as lithium-polymer or lithium-polymer (ab- short-rated Li-Poly or LiPo) has the following characteristic: the lithium salt electrolyte is not contained in an organic solvent, but is found in a solid polymer composite, such as polyacrylonitrile. There are many advantages to this type of construction, including the fact that the solid polymer does not evaporate and is not flammable. Current polymer cells have a flexible, often foldable (polymer laminate) sheet structure, so they do not have a rigid container. They typically have an envelope structure.

Other batteries under study

The development and evolution of batteries has so far been supported by the large consumer electronics market. There are still theoretical possibilities to increase the performance of batteries dozens of times over, and many ongoing studies are currently driven by the prospect of an imminent wide opening of the market due to the spread of electric traction.

• Zinc cells - air: the discharge is obtained by the oxidation of zinc by the oxygen in the air. They use a catalyst to allow the reverse process. They allow high energy densities but low powers

• Lithium - Sulphur cells: they promise high capacities, but the problem of lithium destruction in the electrolyte is not yet solved.

• Lithium - Air Cells: they have a very high potential, but are still being tested for a membrane permeable to oxygen but not to water.

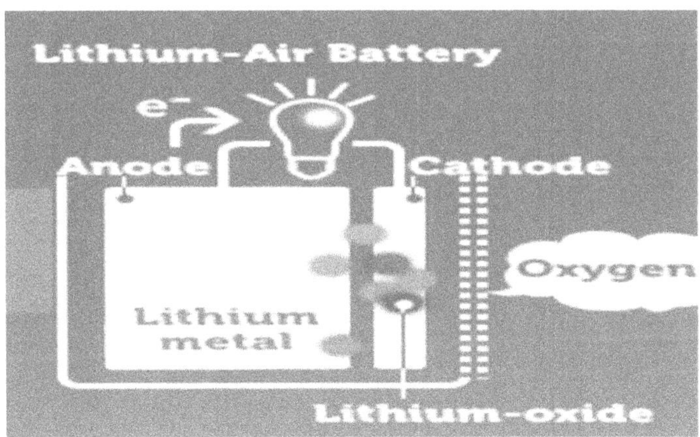

Flow Batteries

In a flow battery two electrolytes flow through the electrochemical cell. The electrolytes are stored externally, usually in tanks, and are pumped through the cell. Flow batteries, when mounted on vehicles, can be quickly recharged by replacing the electrolytes.

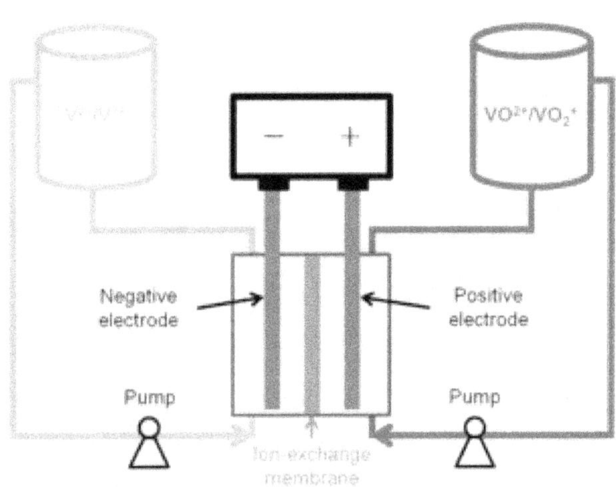

CHAPTER 4 - Battery Management System (BMS)

The "Battery Management System" and abbreviated "BMS" is an essential component of lithium batteries with more than one cell.

It is used to monitor the state of the battery, taking measures and if necessary, intervening, in order to achieve perfect safety and performance optimization.

The results achieved by BMS are as follows:

Temperature control: even if they are now very rare cases, it can happen that the battery heats up more than necessary and emits smoke or even catches fire. The BMS measures the temperature of each individual cell and intervenes by detaching the entire battery from the user, in case a temperature exceeds the permitted limit.

Minimum voltage control: in order to avoid damage due to over-discharge, the BMS isolates the battery if the discharge has reached the permitted limit.

Maximum voltage check: When recharging the battery op-pure, if it is an electric car battery, during regenerative braking, the voltage may rise above the permitted limit. The BMS prevents damage and heating due to this cause.

Current control: The BMS does not allow unsupportable currents from the battery.

Recharging control: the BMS performs a series of operations during recharging. It warns the driver of the amount of charge input, the time still required, etc.

Equalisation: in a battery most of the cells are connected in series, so it is important that none of them is more stressed or damaged than the others. The BMS constantly checks the equal distribution of the charge between all the cells and intervenes, especially during the charging phase, to level the characteristics so that, at rest, all the cells have the same voltage. This operation considerably increases the performance and life of the battery.

Charge status calculation: the BMS provides the amount of charge available in the battery at any time. This data is essential to calculate the number of kilometres the car can still drive.

Calculation of the number of cycles carried out: BMS builds up a historical archive in which it keeps track of the battery's ageing, maintenance, repairs etc.

Communication between battery and vehicle: the BMS enters all the data at its disposal into the CAN on-board communication system, which is in the possession of all modern vehicles. The on-board computer can then display the information on the dashboard.

CHAPTER 5 - Recharging Batteries

The batteries are recharged by an electromechanical component - the battery charger - able to adapt the available mains current to the needs of the battery.

The operations to be carried out by the battery charger are:

- voltage adjustment (generally a lowering);
- conversion from alternating to continuous,
- electricity supply according to pre-established phases.

These operations must be carried out with the highest possible efficiency, without disturbing the electricity grid, safely and automatically.

Charging operations: the BMS carries out a series of operations during the charging phase. Typically: a constant current phase equal to about 1/10 of the capacity value impressed on the container with the initials C10; when the maximum voltage is reached, a constant voltage and decreasing current phase follows. This is followed by repeated rest phases, interspersed with system activation phases, with control that the electrical parameters are within limits.

Trickle charge: for long periods (weeks) of non-operation, the battery charger performs a stand-by charge in which, at fairly long intervals, it checks the state of charge and, if necessary, restores conditions with a small charge.

Rapid charge: in almost all lithium batteries it is possible, provided the charger is capable, to perform a rapid charge that restores almost the entire charge, in times ranging from one hour to a minimum of ten minutes. This operation is important for electric vehicles, and is typically used for Lithium-Iron Phosphate cells. Fast charging requires demanding equipment on the ground: for example, to charge energy for 20 kWh in a car in one hour, it requires an installed power of 20 kW, which is much higher than the average household availability of 3 kW.

Inductive charging: almost all charging stations use the system with battery charger, electrical conductor, plug and socket to bring electricity to the car. This system is called "conductive". However, although not yet commercially widespread, there is 'inductive' charging, without contact between the battery charger and the car. The electrical energy passes from the dispenser to the battery in the form of electromagnetic waves of the "microwave" type. This system, although very convenient, requires perfect positioning of the vehicle, a very short distance between the dispenser and receiver, and an appropriate system on board the vehicle.

Replacing the battery: a system used, although not very widespread, to perform a fast recharge is the replacement of the flat battery with a previously charged one. The method requires the vehicle to be prepared to replace the battery, and a technical-commercial organisation has been set up to replace and recharge the delivered battery on the ground.

Battery charger on board or on the ground: the ground charger (not part of the car) makes it possible not to have the size and weight of the object on the vehicle. Having no space problems, it can be better in terms of efficiency, and can be used by several users. On the other hand, the ground charger requires the vehicle to return to the charging station for sure. It is therefore suitable for fleets that run on a pre-established circuit.

The on-board charger allows greater freedom of movement, as charging can be done at any point on the grid. One method being studied is the use of a part of the traction drive (the inverter) also as a battery charger.

Charging columns: If you have a box or private space with electricity supply, charging can be done privately, using the 220 V mains. Charging points are spreading, e.g. in workplace car parks, supermarkets, exchange car parks, etc. There are also public charging stations.

Whenever a non-private charging station is accessed, compatibility of power sockets is of course necessary. Public stations are equipped with user recognition and energy accounting systems to enable payment.

CHAPTER 6 - How to Regenerate a Battery

Now, we will see together how to regenerate a battery through a rather simple procedure that, if there are no real hardware problems with the battery, should allow you to solve the situation (or at least improve it a bit) without forcing you to spend money.

Having said that, I want to make one thing clear from the beginning: if, after performing the regeneration procedure that I'm about to show you, you don't notice any improvements (small or big), it's likely that you will have to replace the battery of your device, as it could be damaged or worn out.

Regenerate your Smartphone and Tablet battery

Let's start this guide by selling up close how to rebuild your smartphone and tablet battery. Below you can find useful tips on what to do, suitable for both Android and iOS devices. Going into more detail, I will explain how to calibrate the battery: a procedure by which you can reset the battery management by the device and, therefore, solve any battery management problems.

Android

In order to regenerate the battery of an Android device, you first need to completely discharge it until the device is turned off. Then, turn your smartphone or tablet back on using the power button and let the battery run out again until the device is turned off and repeat the procedure until the battery is almost completely depleted (when the device struggles to turn on again).

At this point, charge the device, wait until it is fully charged and, once it reaches 100%, leave it for another couple of hours to charge. Finally, take the battery out of the battery compartment (or keep the device off, if you can't take the battery out) and keep the device "at rest" for about twenty minutes. Then insert the battery back into place and turn the device back on. If everything went well, you will have completed the battery calibration operation in the correct way and you should notice some more or less significant improvements.

If not, you can opt for an alternative procedure, which is to use battery calibration apps (e.g. Battery Calibration Pro), which usually only work on root devices.

How to calibrate the battery on Android using apps

If the above steps have not had any effect, you can try to bring things back to normal by using a specific app for the purpose, such as Easy Battery Calibration: the latter, available free of charge from the Play Store, allows you to restore the correct "interpretation" of the battery status by the operating system. In order for the app to complete its task successfully, it is necessary that root permissions have been previously unlocked on the device in question; otherwise, the result may not be satisfactory.

However, once you have installed and started the app, proceed as follows: connect your smartphone or tablet to the charger until the remaining charge level reaches at least 95%. When this happens, tap on the Calibrate Battery button, grant root access permissions to the app by answering yes to the warning screen shown below and wait a few seconds for the procedure to be completed: from this point on, the battery status should be read successfully again. That's it!

How to calibrate the Android battery without root

If you're looking for a way to calibrate the Android battery without root because you don't intend to "unlock" your phone (or tablet), you can give Battery Repair Life PRO, a free application available on the Play Store that allows, in a single tap, to analyse and restore the "reading" of the charge status of individual battery cells.

Are you asking me how to use it? Don't worry, it's very simple. Once you've installed and launched the app on your device, tap to enter and wait a few moments for the main screen to appear.

Now tap the Start button to perform the first battery cell analysis and wait a few moments for a battery status report to appear. If the app detects problems (inactive or not fully functional cells), tap the Fix Problems! button to try to re-calibrate them. If all is well, the battery reading problems should resolve.

Another important feature of Battery Repair Life PRO is the ability to view information such as voltage, health, temperature, charge status and battery technology at any time: you can access it simply by tapping on the Information tab at the top.

iOS

If you're using an iPhone or iPad, you can regenerate the battery by following a similar procedure to the one I mentioned earlier for Android. In order to do this, even in this case, you need to completely discharge the battery of your device: if the device should turn off before reaching 1% charge, recharge it for a few minutes and then repeat the operation.

When you have finally managed to discharge the device properly, charge it, wait for it to reach the maximum level (100%) and keep it charged for another couple of hours. At this point, restart your iPhone or iPad by holding down the Home button and the phone's power button at the same time (on iPhone X and later, iPhone 8/8 Plus and iPad Pro 2018 and later, you must press the Volume + and Volume - buttons in quick succession and then you must press the side power button) until the apple logo appears.

Regenerate the Laptop Battery

If you want to try to regenerate your laptop battery, you have to follow a procedure that is not that different from the one I indicated in the chapter dedicated to mobile devices. Follow the instructions I'm about to give you and you'll see that you'll be able to get it done without any problems, whether you have a Windows laptop or an Apple MacBook.

The first thing you need to do, again, is to leave the computer on without power, completely draining the battery (of course you can use it to speed it up). But be careful: by completely depleting the battery, I mean that you should also ignore the messages from the operating system warning you that the battery is low and let the notebook go into standby by itself.

Once the computer has gone into standby, leave it running for a few more hours (even a full day if necessary) and wait for the remaining energy in the battery to run out completely and then turn off the computer completely.

When the computer is off, plug in the power supply again and let the battery charge reach 100% (if possible, leave the computer for a few more hours even after the maximum charge level has been reached). Repeat the entire operation three or four times and, if everything goes well, you should notice a marked increase in the battery life of your laptop.

Please do not attempt to increase the battery life of your old battery by unlikely methods that could lead to unpredictable damage. Even if you read that it works, don't put your computer battery in the freezer and don't try to open it by DIY means (unless you are an expert in the field, of course). You'll only risk hurting yourself and killing the battery instead of regenerating it.

What to do in case of further problems

You tried everything but obviously the battery regeneration procedure didn't give the expected results. If this is the case there is a high probability that the battery of your smartphone/tablet or laptop will need to be replaced due to problems that unfortunately cannot be solved by a simple calibration.

If your smartphone, tablet or laptop has a removable battery, you can purchase a new one from electronics stores or online stores such as Amazon. Usually, prices to buy a battery range between $20 and $150, depending on the brand and model of your device.

Before you buy a battery, make sure you know exactly which one is right for your device. Pay attention, then, to the model and brand name of your smartphone, tablet or laptop and check the power (expressed in mAh) of the battery itself, which should be indicated on its label.

If the device affected by the problem has a non-removable battery (which is quite likely, cspccially if it is a latest generation device), to replace it, it is advisable to contact the support service offered by the manufacturer or, at least, a specialized store.

As far as the official support service offered by your manufacturer is concerned, I suggest you visit its website, which should contain all the details on how to request assistance for the problems you have encountered.

CHAPTER 7 – Lithium Batteries (Lipo, life, Lixx)

What the numbers on the battery indicate

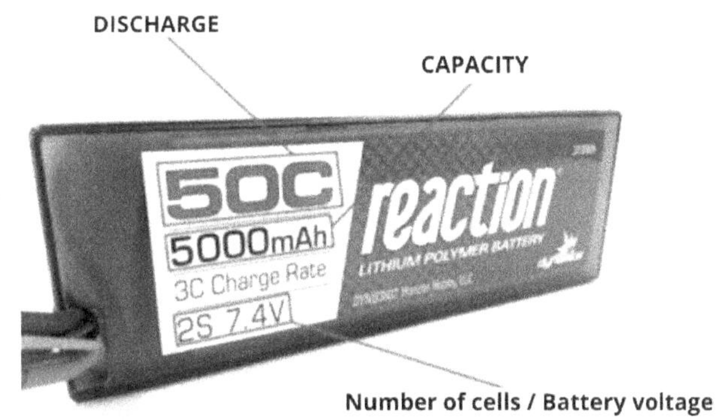

Lipo batteries are composed of cells. The numbers 1S, 2S, 3S... indicates the number of cells that make up our "battery pack".

A single LiPo cell has a nominal voltage of 3.7 V. If our example battery pack has 7.4V, it means that there are two cells in series (The n-cell voltage is added together). This is sometimes why you will hear about a "2S" battery pack - it means there are 2 cells in series. So, a two-cell pack (2S) is 7.4 V, a three-cell pack (3S) is 11.1 V, and so on.

The nominal voltage is the default resting voltage of a battery pack. This is how the battery industry decided to discuss and compare batteries. It is not, however, the full charge voltage of the cell. LiPo batteries are fully charged when they reach 4.2 v/cell and their minimum safe voltage, as we'll see later, is 3.0 v/cell. 3.7 v is practically in the middle, and that's the rated cell voltage. When choosing a battery, it is considered that the average voltage will be the nominal one and therefore the components will be sized accordingly.

You may find batteries with the indication "2S2P". This means that there were four cells in the battery; two cells wired in series and another two wired in the first two batteries in parallel (capacities are added in parallel). This terminology is not used much nowadays; modern technology allows us to have the individual cells with much more energy than was available only a few years ago.

Capacity of a Lipo battery (stored charge)

The capacity of a battery is expressed in mAh (or in Ah for very large batteries).

The "**A**" indicates amperes, a measure of current. (the number of electrons flowing through our power cable every second) It should not be confused with voltage, which is a measure of electrical potential. Voltage and current are linked by Ohm's law.

The "**mA**" stands for milliampere, the thousandths of an ampere.

The "**h**" indicates the time as a unit of time.

Example: suppose we have a 4000mAh battery. Suppose we draw 1000mAh continuously, the battery charge will last 4000/1000 = 4 hours.

So, the capacity of a battery gives me an indication of how much energy the battery can deliver and for how long I will be able to power my load.

Discharge

A typical indication of Lipo batteries is the "Discharge rate", the speed at which I can discharge it safely and without damaging the battery. This parameter is very important, especially if we are going to use brushless motors where peak currents can be very intense. The discharge rate is indicated in C, for example 55C. C in this case stands for "Capacity". Let's make an example: let's suppose we have a 2S 7,4V 5000mAh 50C battery The 50C indicates that we can draw a maximum current of 50 x 5000mA = 250A That means our battery will not be ruined if we are going to draw very intense currents. Obviously, the duration of the charge will be very minimal. If we draw that current continuously, our battery will be discharged after about one minute. (theoretical)

Charging current

Another indication that may be present on the battery label is the charging current. Generally, Lipo batteries can be charged at a current of 1C. As we have seen before, C indicates the capacity of the battery, so the maximum current we can use for charging will have to be evaluated on a case by case basis. Let's make an example: suppose we have a 2S 7.4V battery 5000mAh 50C in this case the charging current 1C will be equal to 5000mA x 1 = 5A. Given the currents in play and the risk of causing serious damage (the batteries could catch fire) we advise you not to use self-built equipment, but to opt for approved and electronically controlled objects. Modern battery chargers have present programs that make it easier to charge, discharge and store batteries. There are chargers that cover various types of batteries (Pb, Lipo, Nimh, Nicd) all in one.

Charge Lipo batteries correctly

It is important to use a compatible LiPo charger. Unlike other batteries, LiPo batteries require "special care. They are charged using a system called a DC / CV charger. It stands for Constant Current / Constant Voltage. Basically, the charger will maintain a constant current, or charge rate, until the battery reaches its peak voltage (4.2V per cell in a battery pack). Then it will maintain that voltage, reducing the current. For example, NiMH and NiCd batteries charge better using a pulse charging method. Charging a Li-Po battery in this manner can have harmful effects, so it is important to have a Li-Po compatible charger.

The second reason you need a Li-Po compatible charger is balance. Balancing is a term we use to describe the act of equalizing the voltage of each cell in a battery pack. We balance LiPo batteries to ensure that each cell discharges the same amount. This helps with battery performance. It's also crucial for safety reasons, but we'll get to that in the discharge section.

While there are independent balancers on the market, I recommend purchasing a charger with built-in balancing features (almost all chargers currently on the market have this feature).

Balancing and charging Lipo batteries

Most LiPo batteries come with a connector called the JST-XH connector on the balance cable. The chargers have dedicated connectors, or small external cards to connect to the balance cable. It is critical that the balancing cable is always connected before charging. Most LiPo batteries need to be charged quite slowly compared to NiMH or NiCd batteries. While we normally charge a 3000mAh to four- or five-amps NiMH battery, a LiPo battery of the same capacity should be charged at no more than three amps. Just as a battery's C Rating determines how safely a battery is discharged, there is also a C Rating for charging. For the vast majority of LiPo, the rate is 1C. The equation works the same way as the previous discharge rating, where 1000mAh = 1A. So, for a 3000mAh battery, we would charge at 3A, for a 5000mAh LiPo, we would set the charger to 5A and for a 4500mAh pack, 4.5A is the correct charging rate.

Using Batteries in Parallel

Using, but also charging batteries in parallel can be extremely dangerous, because the batteries may catch fire. Cell voltages may not be properly balanced with each other. So, the batteries are not identical, but current may flow from one battery to another. These currents can also be very strong and may cause the batteries to overheat. In the paragraph "Safety" we will see that the ignition temperature is quite low. For this reason, we recommend never leaving a Lipo charger unattended, and if necessary, use a fireproof safety bag.

Lithium battery safety

It has been proven that, once the 75 °C threshold has been exceeded, the internal overheating reaction becomes so autonomous that it generates a non-return phenomenon defined as Domino Effect in which, although acting on the battery, it is impossible to avoid the explosion.

To avoid damage or unpleasant accidents, never charge batteries placed on flammable surfaces.

If you do not have the ability to monitor charging or avoid potentially flammable surfaces, you can use safety bags.

LiPo, LiIon, LiFe, LiHv batteries: charging, discharging and handling.

<u>Never Treat Li-Po Or Li-Ion Batteries In The Same Way As Other Battery Types!</u> Li-Po and Li-Ion cells are much more sensitive and volatile than NiCd, NiMH and Pb batteries. Improper use or overcharging can cause the lithium cells to overheat and swell, which can quickly lead to <u>violent explosion and/or fire</u> that can cause serious damage and injury. <u>Never</u> leave lithium batteries unattended while charging or discharging! <u>Never</u> place lithium batteries on flammable surfaces while charging. Never attempt to use NiCd, NiMH or Pb charging functions for lithium batteries. Never use lithium batteries that have begun to swell or overheat easily or that do not charge sufficiently in the specified time. Failure to do so could result in an explosion or fire!!! NEVER discharge lithium-ion cells below 2.5V per cell.

Before we begin with the theory of charging lithium-ion batteries, we would like to point out that an optimal process is achieved with the microprocessor technology now integrated in good quality chargers. An example of a computer charger that "charges everything" is the IMAX B6.

Lithium cells do not suffer from self-discharge and memory effect; therefore, they should not be subjected to charge/discharge cycles. Even if the idea remains, deriving from old generation NiMh batteries, contrary to what is thought, lithium batteries should be charged "as soon as possible", whenever a socket or charger is available. Long charge and discharge cycles significantly reduce battery life. If you are using batteries from your mobile phone or a portable gadget, we recommend a full charge and discharge once a month. If you are using a microprocessor charger, it will have several modes: full charge, regeneration, storage. We recommend that you refer to the instruction manual for correct use.

Lithium cells must be fully charged at least once a year to prevent under-discharge. Lithium batteries should be stored with about 30-50% of their capacity if they are not used for a long time. If the battery becomes damaged or swollen <u>do not use it!!!</u> Replace it immediately with a new, undamaged one. There are three types of Lithium batteries: Lithium Polymer (LiPo), Lithium Ion (LiIon) and Lithium Iron Polymer (LiFePo4). The most used in modelling are definitely the LiPo for their incredible efficiency and power and the LiFe for their reliability and robustness. Due to their high potential Lithium batteries need special and careful handling as they can be easily damaged and even <u>explore!!!</u> So always handle your Lithium batteries with extreme care and follow your battery specifications scrupulously and always check the charger settings. Despite their apparent delicacy, Lithium batteries are actually much more efficient and, due to their chemistry, do not require any special cycles or processes to maintain them at their best. They do not suffer from the memory effect and have a very low self-discharge so management is much easier. The only shrewdness that you must have, in addition to using them within their specifications, is to always keep the cells inside the pack always balanced. In this regard, the Equilibrium battery chargers have a dedicated and optimized program for balancing the cells, provided you connect the balancing socket on your battery pack. Through the balancing connector, in fact, the charger checks the voltage of each single element of the pack by applying small loads (about 200mA) to keep at the same voltage.

Charge Lixx batteries. Two different programs are available to charge Lithium batteries, one dedicated purely to charging (CHARGE) and the other dedicated to balanced charging (BAL-CHG). In both processes it is good to connect the balancing connector, however in the simple charging process less attention is paid to balancing and the tolerance is much higher. This makes the charging process faster but is strongly discouraged unless you have a perfectly balanced and new pack per se, in which case if you are in a hurry you can use the CHARGE process to have the battery ready as soon as possible. Do not abuse it, and always check that the package remains balanced before using it, if you see that the package tends to unbalance it is good to use the BAL-

CHG program. The balanced charge in fact pays extreme attention to the balancing of the individual cells and until it has completely balanced all the cells, the process does not end. For this reason, completing this process takes considerably longer than a normal charge, but the balancing of the battery is guaranteed. It is strongly recommended to always carry out this process and to prefer it to normal charging for safety and performance reasons. General notes regarding the charging of lithium batteries: Always observe the charging current values indicated in the battery specifications. If a battery is charged at 1C do not charge it at higher currents! It can be irreparably damaged. Lithium batteries are charged using the "constant current / constant voltage" process. Constant current is delivered in the first part of the fast charging process. When the battery reaches the present voltage, constant current will no longer be supplied and constant voltage will be applied to the pack. When the battery voltage becomes equal to the output voltage of the charger, the charging current will start to drop immediately. This is normal. When the current reaches an approximate value of 1/10C, the charging process will be completely finished. LiPo/Ion batteries do not require compensating charging and there is no function in the charger For lithium batteries 11.1V or more, the current amount of current supplied to the battery may be reduced due to the charger's power limitation. This is normal. For safety reasons, all cut-offs should be set for capacity and time. The backup safety timer can be set to provide an additional level of safety. If the battery pack is extremely unbalanced or the charger fails to balance the battery pack correctly it may mean that one or more cells in the pack are damaged. Do not use the pack, replace it with a new one, using an unbalanced pack or one that tends to become very unbalanced can be dangerous! Lithium battery storage process If you do not use the Lithium batteries for a long time it is good to perform an Equilibrium process called STORAGE before storing them. Through the STORAGE process, the Equilibrium brings the battery to the storage voltage which prevents the Lithium cell from deteriorating. It is strongly recommended to perform this process if you do not use the batteries for a long time (maybe a month), otherwise the battery will be damaged and its performance will decrease considerably. The recommended storage voltage is just above the rated voltage (around 3.8V for LiPo), however the STORAGE program will automatically bring the battery pack to the recommended voltage based on the data entered. The STORAGE process will then load or unload the battery pack as needed. It is strongly recommended to use the balancing connector to ensure a more accurate result. Discharging Lithium batteries Unlike NiMH and NiCd batteries, discharging Lithium batteries is of little use. Charging a Lithium battery when it is already half charged has exactly the same effect as charging a fully discharged battery, unlike NiXX batteries. In any case, if you need to discharge your batteries, the DCHG program is for you. The Equilibrium chargers do not discharge batteries below a certain voltage (3.0V for LiPo and LiIon and 2.0V for LiFe), however, you can adjust this value upwards if necessary. Also, during the discharge process, it is advisable to connect the balancing socket to control the behaviour of the individual cells and balance the pack.

CHAPTER 8 – How to Build a Lithium Battery Charger

In this chapter we will make a battery charger that allows us to charge lithium Lipo.

In fact, as we know, they need a special charger that allows you to charge them correctly without exploding or ruining them.

Cell Charge

The charging current must be less than the maximum charging current specified in the technical specifications. Logically higher current charges will permanently damage the battery performance.

Even the charging voltage must not exceed that indicated in the technical specifications, because the higher values also damage the battery performance and violate the safety characteristics of the elements, very often they can also lead to an explosion.

The same goes for the polarity, in fact, not respecting it you risk to explode the battery, especially during the charging cycle.

How It Works

The battery charger is a fundamental element to use these batteries, in fact these batteries can NOT be charged with a normal battery charger, but with a particular charger capable of delivering a voltage and a current capable of not exploding the batteries.

We must know that these batteries up to 70% of their charge, the voltage charge must be at maximum to be able to charge them, then as the battery begins to reach its full charge, the output voltage of the charger must decrease until it becomes null.

Construction

The scheme I used is the most common on the internet, it works very well and charges the batteries perfectly in a couple of hours.

On X2-1 enters the voltage of 15v able to charge the batteries... the voltage regulator LM317 regulates the output voltage that according to the cells present will be with a cut-off voltage of (4.20±0.05) Volt or for 2 cells of (8.40±0.05) Volt and to finish with 3 cells (12.60±0.05) Volt.

All the resistors present, from R9 to R16, are used for the output current which, depending on the quantity of cells present in series, increase or decrease the output current.

The Circuit

As we can see, the circuit here unravelled takes up little space, allows to charge up to 3 cells in parallel with output amperages of 985mA.

From the measures of 6.5 x 4.5 cm uses components easily recoverable in any electronics store, with standard pitch, its cost is on 10$.

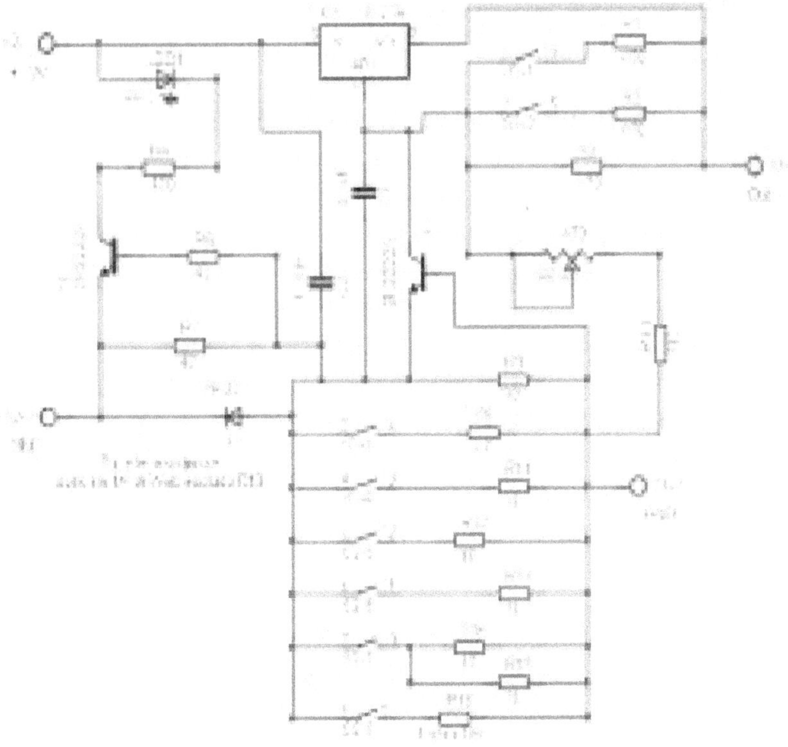

The list of components required for the realization

Resistances	
R1-R4	470Ω 5% ¼ Watt
R5	Linear Trimmer 470Ω

R6-R9	47Ω 5% ¼Watt
R10	1KΩ 5% ¼Watt
R11	10Ω 5% ¼Watt
R16	1Ω 5% 1Watt

Capacitors	
C1-C2	0,1µF Polyester
Integrated	
U1	LM317T
Diode	

D1	1N4001
DS1	LED Red
Semiconductor	
Q1-Q2	2N2222A

Other	
Heat sink	
Switches	DIP16, 8 Switches
X1-X2	2 seater connectors
Box	To contain the battery charger

The adjustment table

Amperage current output regulation table

S3	S4	S5	S6	S7	S8	
OFF	OFF	OFF	OFF	OFF	OFF	10 mA
ON	OFF	OFF	OFF	OFF	OFF	24 mA
OFF	ON	OFF	OFF	OFF	OFF	62 mA
ON	ON	OFF	OFF	OFF	OFF	75 mA
OFF	ON	ON	OFF	OFF	OFF	127 mA
OFF	ON	ON	ON	OFF	OFF	192 mA

OFF	OFF	ON	ON	ON	OFF	260 mA
ON	ON	ON	ON	ON	OFF	335 mA
OFF	OFF	OFF	OFF	OFF	ON	650 mA
OFF	ON	OFF	OFF	OFF	ON	725 mA
OFF	ON	ON	OFF	OFF	ON	790 mA
OFF	OFF	ON	ON	ON	ON	907 mA
OFF	ON	ON	ON	ON	ON	972 mA

ON	ON	ON	ON	ON	ON	985 mA

This other table is used to quantify the cells that will be connected to the charger.

S1	S2	
OFF	OFF	1 Cell = 4.2v
ON	OFF	2 Cells = 8.4v
ON	ON	3 Cells = 12.6v

For example, the configuration for a pack of 2 cells connected in series is 1100mAh is:

- S1 ON
- S2 OFF
- S3 OFF
- S4 OFF
- S5 OFF
- S6 OFF
- S7 OFF
- S8 ON

So, 8.4v and 650mA output

Thanks to this charger we will be able to charge lithium polymer batteries without any problems during charging.

Always remember that it is recommended to adjust the charger before charging the batteries and thus avoid problems during the charging cycle.

CHAPTER 9 - Lithium Batteries: How to make them last longer

The battery is so fundamental that part of companies' efforts is linked to its use and to the extension of battery life itself: today, in fact, the usability of many products, first of all smartphones, is linked to the inability of a battery to accumulate current. Although lithium polymer models are among those with the highest current density compared to the size of the battery, this is not enough to solve many problems today. Everyone is looking for solutions, but lithium technology is the most tried and tested and we'll keep it tight for a while.

The battery is obviously a current magazine, and its capacity has a unit of measurement, the milliampere now, commonly written mAh. This indication, found on the back of battery packs and most smartphone batteries, indicates how many milliamperes, or how much current a battery can deliver in an hour.

From the number of mAh in any case is very difficult to trace the autonomy of the device, because it is the latter that "asks" current to the battery: for example, in the computer this request varies depending on what the computer has to do. You are converting a movie, work that requires power from the CPU, the current consumption will obviously be much higher than a web browsing situation. In general, however, the greater the milliampere now, the greater the capacity of a battery to last over time.

Discharge and Charge Cycles

A lithium battery has a life cycle that is not very long: you can safely say that all products contain batteries that have a duration ranging from 200 to 500 charge/discharge cycles, depending on the goodness of the battery. Inside each lithium cell a series of chemical processes take place, and the deterioration of the materials due to continuous charging and discharging leads to a reduction in capacity over time. If a good battery after 450 cycles can reach about 80% of its original charge, a bad cell can also halve its charge.

The most important thing in any case is never to let a battery completely discharge: lithium batteries do not suffer from memory effect, this is true, but due to their nature they should never be completely discharged. Paradoxically, and this goes against every common belief, to make a battery last longer it should be charged very often, as soon as possible. A smartphone lives much longer if when we are at home, we recharge the battery even if it is 50%, the shorter the charge/discharge cycles the longer the battery lasts. You don't have to worry at all about continuing to disconnect and attach your smartphone to the charger: a partial charge, even 20%, doesn't count as a full charge/discharge cycle.

When you don't use a device for a while the best thing to do, before putting it away, is to check that the battery is at least 50% charged: a lithium battery, as we have already said, is ruined if the internal cells fall below a certain threshold and a lithium battery (technology that can hold the charge well for several months) 50% charged gives us good security.

Periodically it is advisable, especially for smartphones and tablets, a complete discharge with 100% charge to "calibrate" the battery: in reality you are not calibrating the battery at all, which is an "ignorant" lithium cell, but you are just adjusting the system that manages the battery for the smartphone. Inside every product with a lithium battery there is a circuit that ensures that the battery never drops below a certain voltage, about 2.7V, because it would be irreparably damaged. The same circuit also makes sure that the cells, if more than one, are all charged in the same way and with the right voltage (which are not the 5V of the USB charger).

Two other elements to take into account, even if only one of them can be affected, are the heat and the charging voltage. Lithium batteries suffer a lot from heat, and already a temperature above 30° is considered high for a lithium cell: it is necessary to keep a battery at room temperature and avoid charging a smartphone that is already hot for other reasons, because it would not be able to hold all the charge.

The battery of a smartphone that has reached a high internal temperature - and here the HTC One M9 with the Snapdragon 810 comes to mind - will lose performance much faster than the battery of a smartphone that, in addition to dissipating heat perfectly, does not even heat up that much.

The other interesting fact is related to charging: a lithium cell lasts much longer if it is charged at a present voltage of 3.92V. At this voltage you can also reach 4000 life cycles, however, it is not possible to reach the full capacity of the battery itself: this means that by charging a 3300 mAh cell at 3.92V this cell will last very long, will remain in full form for several years but will be able to deliver, when charged, at most 2000 mAh.

Electronics manufacturers obviously can't do this reasoning, they need the maximum capacity, and they raise the charging voltage to about 4.2V: in this way the battery life starts to suffer after a year or so, but you get 100% charging capacity.

Exaggerated charging voltage also leads to the dangerous phenomena of swollen batteries, with the risk that they may burst (and this has already happened). Fast charging is not too harmful, even if it is not very good for batteries: it is a good strength to show off for smartphone manufacturers and does not reduce the life of the product, as long as it is managed by a dedicated charger with a control processor as in the case of Qualcomm's Quickcharge technology.

Among other myths to be debunked there is also the one related to the need to disconnect the charger from the smartphone when the battery is charged: the circuit that manages, inside the smartphone, the charge of the cell is smart enough to be able to stop the charge at the right time.

CHAPTER 10 – Safety

In this chapter we will see the 3 key factors in the safety of lithium batteries:

1. Choosing the right chemistry for lithium batteries
2. Type of assembly of lithium batteries
3. Electronics controlling the lithium battery

Choosing the right chemistry

There are hundreds of different chemicals for lithium batteries on the market, but the most used ones are 3:

- Lithium NMC - Lithium Nikel Manganese Cobalt (LiNiMnCoO2)
- Lithium NCA - Lithium Nikel Cobalt Alluminum (LiNiCoAIO2)
- Lithium LFP - Lithium Iron Phosphate (LiFePO4)

Our goal is to identify the safest lithium chemistry for your vehicle in order to avoid unpleasant risks when using lithium batteries.

	Safety	Decomposition temperature	Heat release
NMC	★ ★ ★	210°C(410°F)	600J/g
LFP	★ ★ ★ ★ ★	270°C(518°F)	200J/g
NCA	★ ★	150°C(302°F)	940J/g

In this diagram we have summarized the elements intrinsic to each chemistry in terms of safety and the values that determine it.

Decomposition temperature: the higher this temperature is, the more difficult it will be to reach the decomposition conditions; therefore, the lithium battery cell will be safer. It refers to a situation where an increase in temperature leads to a condition where there is a further increase in temperature, for example a domino effect between the cells inside the battery, also called Thermal Runaway.

Heat released: is measured in Joules per gram and indicates the energy that can release the battery cell in the form of heat, thus contributing to the temperature rise. The lower this value, the safer the lithium battery will be. It is precisely the increase in temperature that causes the most damage, the fire may not come directly from the cell but may ignite on components near or around the battery, for example plastic material.

Nail Penetration Test

The Nail Penetration Tests are an excellent example of safety tests on lithium batteries performed to simulate the internal short circuits of the cells and see how they evolve.

In this way the perforation simulates an internal short circuit and allows you to verify that the battery does not catch fire or explode.

The probability that a cell is penetrated into the common use in a lithium battery installed on electric vehicles and industrial machinery is very low, but we take this test because it simulates the worst thing that can happen to a cell, the internal short circuit that can occur due to any manufacturing defects or abuse.

Type of assembly of lithium batteries

The assembly is another fundamental point from which the intrinsic safety of the battery passes, specifically the number of parallels inside the pack is the decisive element.

Remaining in the world of lithium batteries for electric vehicles and industrial machinery the needs are those of having an average high capacity that can go from 100Ah over 1000Ah.

Is the safety of lithium-LFP maintained in all conditions?

Many lithium battery manufacturers assemble battery packs with small cell sizes so they are forced to put a very large number of cells in parallel.

Let's think of a 400Ah battery.

If it is composed of 3Ah cylindrical cells, we would need 130 cells in parallel, if it is composed of 50Ah prismatic cells, in this case we will have 8 cells in parallel.

What happens if one of these 8 or 130 cells should short-circuit?

The lithium cell that will short-circuit will have to absorb all the energy of the entire parallel or 8 or 130 times its capacity.

In this way the temperature of the cell will increase exponentially, putting the safety of the entire battery or the entire vehicle at risk.

Control electronics in the lithium battery

The third and final aspect to ensure the safety of the internal lithium battery pack is the electronics that controls the battery, the brain of the battery.

The main task of the electronics is to monitor the voltage and temperature of the individual cells, in addition to this must communicate with the vehicle and battery charger to stop charging and discharging in case of critical situations and possibly intervene on the general contactors.

The difference between lithium battery manufacturers is how the control electronics work in hazardous situations.

Traditional systems monitor the temperature every 3-4 cells and not always in the right places.

CHAPTER 11 - How to Build an Electric Bike Battery Pack

Requirements

The battery pack in question must supply power to a DC/DC converter with the following characteristics:

- Nominal supply voltage: 36Vdc.
- Total power: 350W.

The design requirements require a Lithium technology battery pack with the following characteristics:

- Nominal voltage: 36V.
- Output current: 12A.
- Total energy: 420W/h.
- Estimated autonomy (average operating conditions): about 2h.

The battery pack must be made using Lithium elements of shape 18650 with a capacity of 1.5 A/h.

Realization

For the realization of a battery pack with the characteristics described above it is necessary to realize a series of strings to be connected in parallel.

Recovery batteries have been used, taken from old laptop batteries.

Each string is composed of a series of 10 batteries; this imposes the voltage at the ends of the battery pack and since the nominal voltage of a single lithium cell is 3.6V, connecting 10 of them in series the final voltage will be 36V (nominal).

The parallel connection of 'n' strings determines the capacity of the entire battery pack since the capacity of the single string (which is equivalent to the capacity of the single battery), is added to the elements that are placed in parallel.

Considering that each battery has a capacity of 1.5A/h, having to supply an output current of 12A, it will be necessary to connect 8 strings (1.5Ax8 = 12A) in parallel.

For the realization of a battery pack with the necessary requirements it is therefore sufficient to follow the model "10s8p" which is composed of 8 strings in parallel; each string is composed of 10 batteries in series.

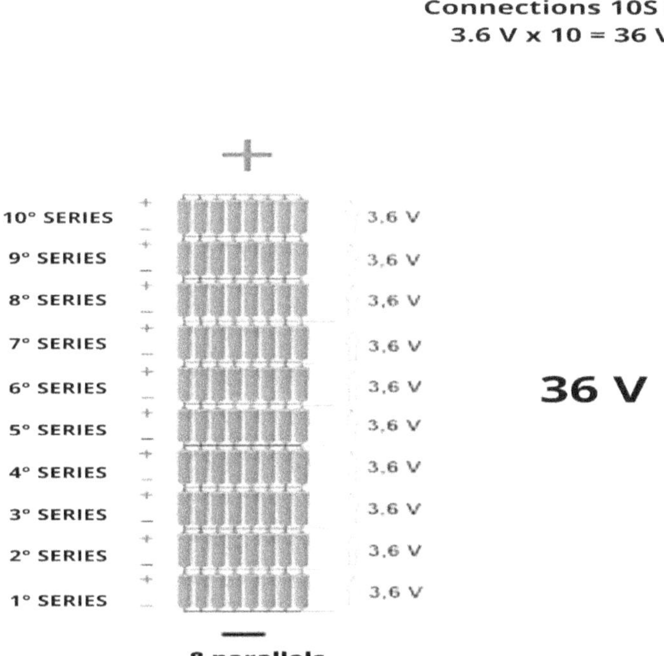

The most delicate aspect of this configuration, which certainly deserves more attention, is the management of the series connection of the batteries.

In fact, the correct functioning of the entire package is subordinate to the functioning of each single cell placed in series;

the malfunction of even one of the batteries placed in series is sufficient to compromise the overall functioning of the entire package.

More specifically, the SOC (State-of-Charge) of the battery that has the lowest level, within a string, compromises the SOC of the entire pack; in fact, this determines the total autonomy of the battery pack to the point of compromising its operation in the most serious cases.

For this reason, it is necessary to balance the SOC of each battery to make it as equal as possible to that of the other batteries in the same string.

The balancing of the SOC of each element is carried out during the charge and discharge of the string through the use of a suitable electronic device.

This device is called BMS (Battery Management-System) and must be connected in parallel to each cell-battery making up the string.

Vbatt (nominal): 36V 10s: 1 string of 10 elements in series -Batt.

In fact, the BMS automatically manages the current flow in/out of each battery reducing it in case of a battery with lower SOC and increasing it in case of a battery with higher SOC.

This balance tends to balance the SOCs of all the elements that make up the string by increasing the SOCs of the battery (in some cases even considerably) the autonomy and average life of the battery pack.

The BMS is equipped with a port for the battery charger input, a port for the power supply output (on which the battery pack provides power) and finally a port on which there are the inputs to be connected to each battery cell.

The various battery cells of each string are connected to each other in parallel so it is sufficient to

connect each "Bn+" line of the BMS to a single battery to balance the same batteries placed in the same position on the other strings.

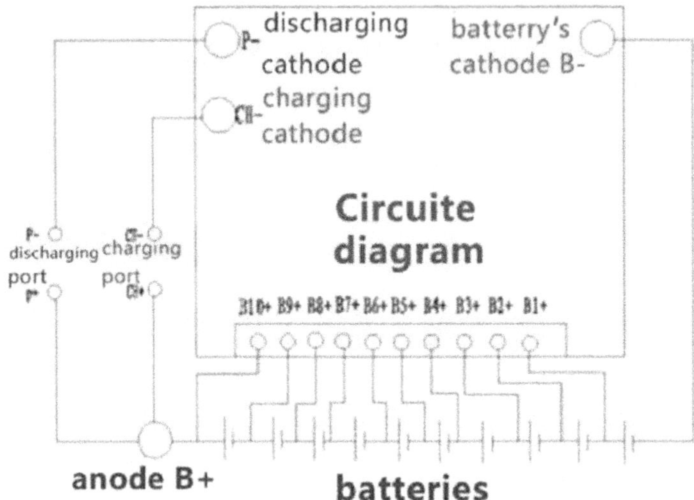

By making parallel connections between the various batteries, each 'parallel' branch of the battery pack will assume the same SOC; by balancing, through the BMS, the SOC of each battery of each string, the battery pack, as a whole, will assume a uniform SOC.

It is therefore of fundamental importance to insert a BMS inside the battery pack; this operation must be done following the diagram shown in the picture.

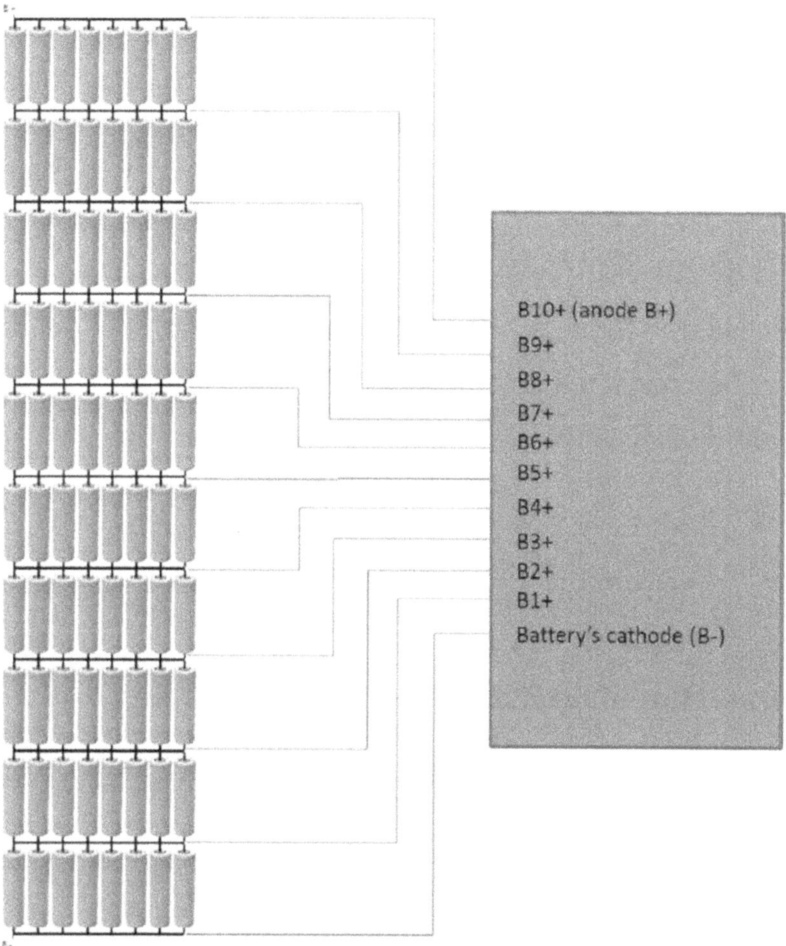

The battery pack charge and power supply of the converter compete at two different ports.

The battery charger is connected to the charging-port.

The charging current must be between 1/10 and 1/2 of the battery capacity.

In our case, since our pack has a capacity of 12 A/h, the charging current can be between 1.2 and 6 A/h.

The load, the converter that will then supply power to the motor, must be connected to the "discharging- port"; all the current required to move the motor at various speeds will circulate on this port.

Connection in series

To make a 10S8P battery pack we have to make:

- 10 series = 36 V
- 8 parallels = 16 Ah

To calculate the current, keeping in mind that each battery delivers a nominal current of 2 A, putting 8 of them in parallel we will have: Total current = 8 batteries x 2 A = 16 Ah

Tools

- Welder
- Pond
- Cutting knife
- Screwdriver
- Screwdrivers
- Scissors

Components list

Batteries size 18650:

- Samsung 30Q
- Sony VTC6
- LG HG2
  - Bms
  - Capacity meter display.
  - Battery holder
  - Fuse Insulators
  - Fuse holder
  - Battery charger (internal)
  - Millefori card

- Battery charger (external)
- Electrical clamps
- Thermo shrinking sheath
- Rubber sheets

Battery recovery

The batteries inside the containers are very well sealed, so it is easy to damage them or hurt yourself in order to extract them intact.

For this reason, use all the appropriate protection and safety devices to avoid accidents or short circuits.

But be very careful when trying to remove the batteries from the plastic containers because they are well bonded and always wear gloves and eye or face protection (everything flies when they crack).

Break it into a large, transparent plastic bag so you can put your hands inside and at the same time see where to "stick" the battery.

Preparation

Before inserting them into the battery pack, you must charge all batteries and discard those that overheat or never charge.

The batteries are charged at a voltage of 4.20 V.

At the end of charging you can read the accumulated capacity of each cell on the display of the portable charger and make a note of it.

CHAPTER 12 – Process

In this chapter you will see the practical steps to create a battery using the 18650 size.

In this case we are talking about a battery with these features:

- Rated voltage: 36VDC
- Continuous current: 16A
- Continuous energy: 570Wh

To make the battery pack I used a cardboard box in order to speed up the cutting and drilling operations, while remaining fully functional.

Subsequently, once assembled the package, you can move on to make a box using other materials: plastic box or sheet metal.

In addition, this battery provides an innovative element compared to the packs commonly found on electric bicycles.

It was deliberately used containers with quick extraction of the 18650 modules to avoid breaking the whole package in case of early deterioration of even just one module.

General outline

This is the general scheme we have to follow to make our 36VDC battery pack.

On the left are the cards containing the 18650 modules:

- Card A: modules from 1 to 4
- Card B: modules from 5 to 8
- Card C: modules from 9 to 10

On the right we see the BMS, the battery charger, the battery disconnects and the control unit that we have to power, that is the control unit that controls the engine of our electric bike.

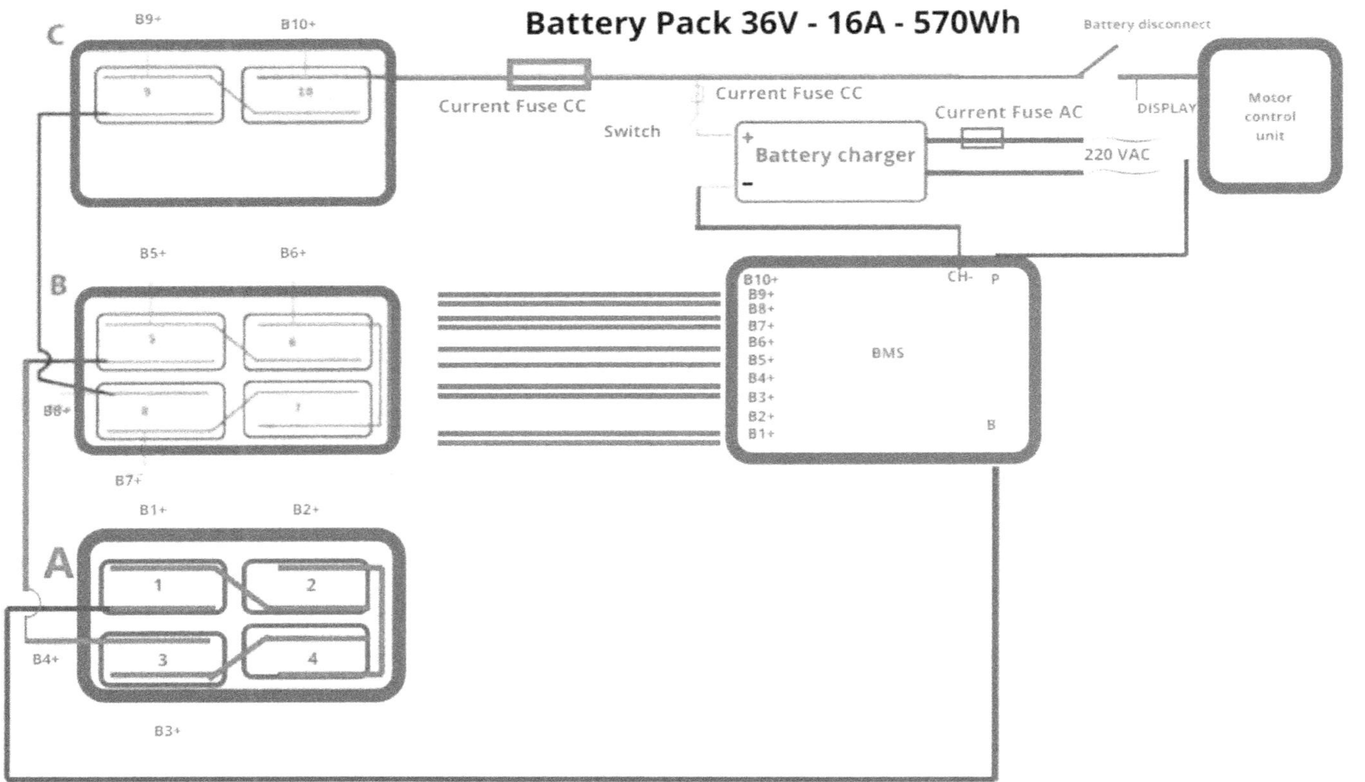
Battery Pack 36V - 16A - 570Wh
Battery disconnect
Motor control unit
DISPLAY
Current Fuse CC
Current Fuse CC
Current Fuse AC
Switch
Battery charger
220 VAC
C
B9+
B10+
B
B5+
B6+
B8+
B7+
A
B1+
B2+
B4+
B3+
1
2
3
4
BMS
B10+
B9+
B8+
B7+
B6+
B5+
B4+
B3+
B2+
B1+
CH-
P
B

From the outside, the battery pack is equipped with the following components:

- Power socket
- Battery charge indicator
- Led battery charger
- Battery disconnector switch
- 36VDC output with external fuse

Outside

Battery disconnect

The battery disconnect switch has the task of disconnecting the battery pack from the engine control unit in the event of a fault.

It is capable of electrically disconnecting a load up to 100 Amps, such as a short circuit or overload.

Moreover, being removable from its housing, it can also be used as an anti-theft device.

Power Cables

The power cables must have a generous cross-section due to the high current flowing through them.

In addition, on the positive side, an external fuse is provided for quick replacement.

Drill a rectangular hole to house the charge indicator on the front.

Internal Components

Dimensioning of Cards

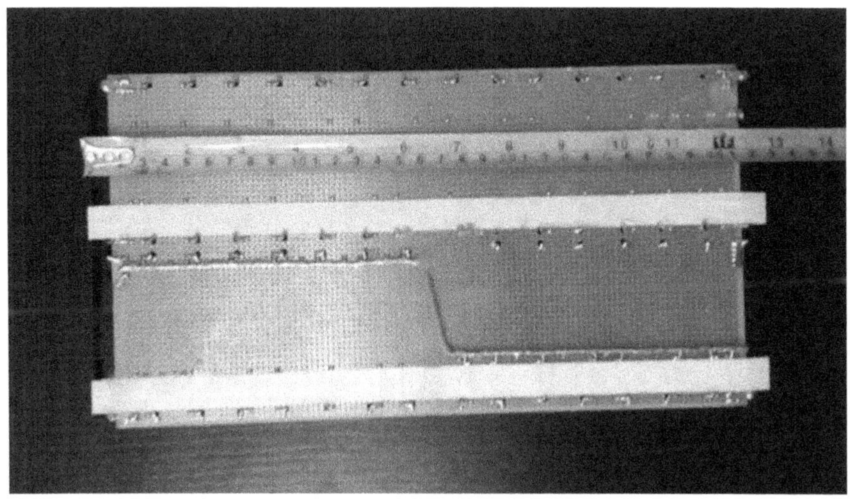

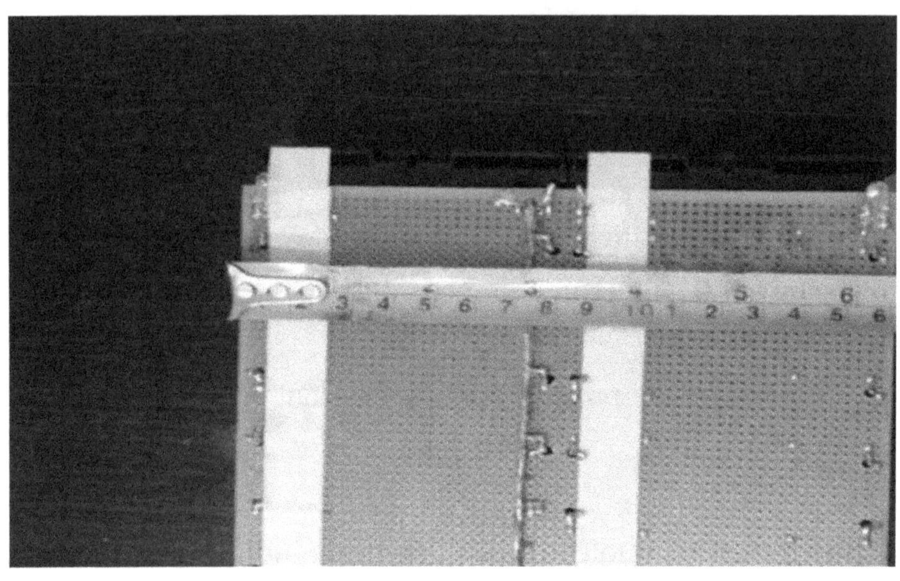

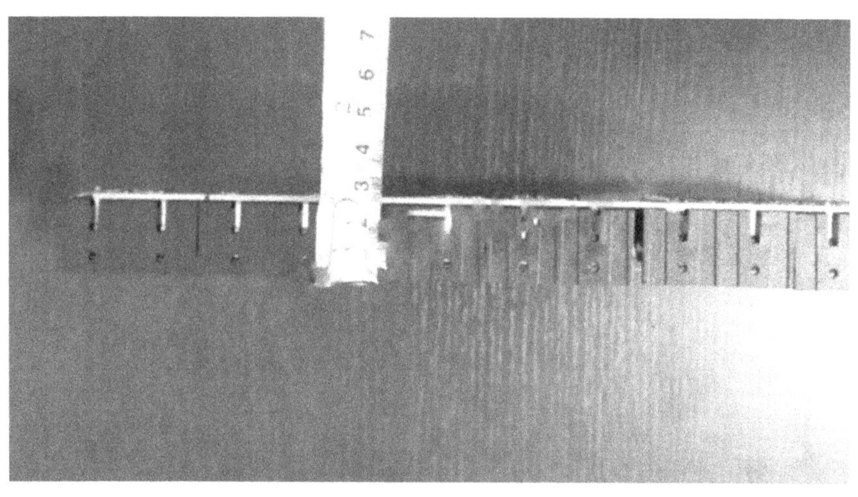

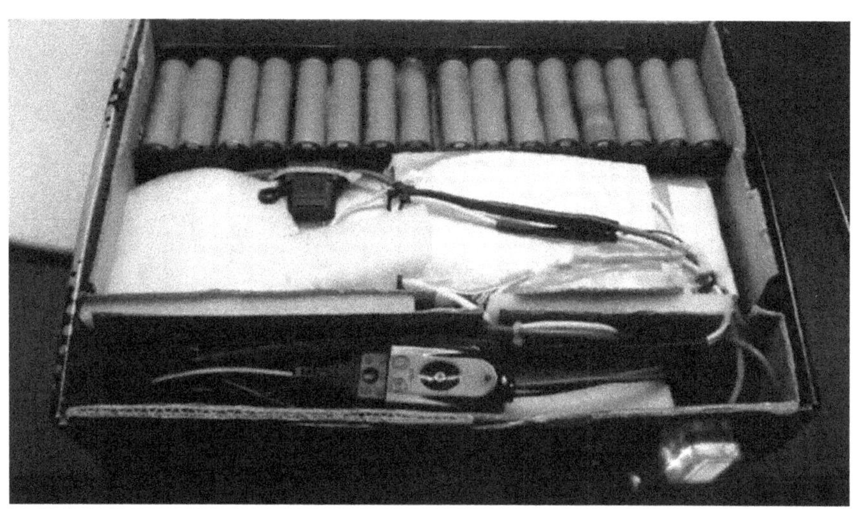

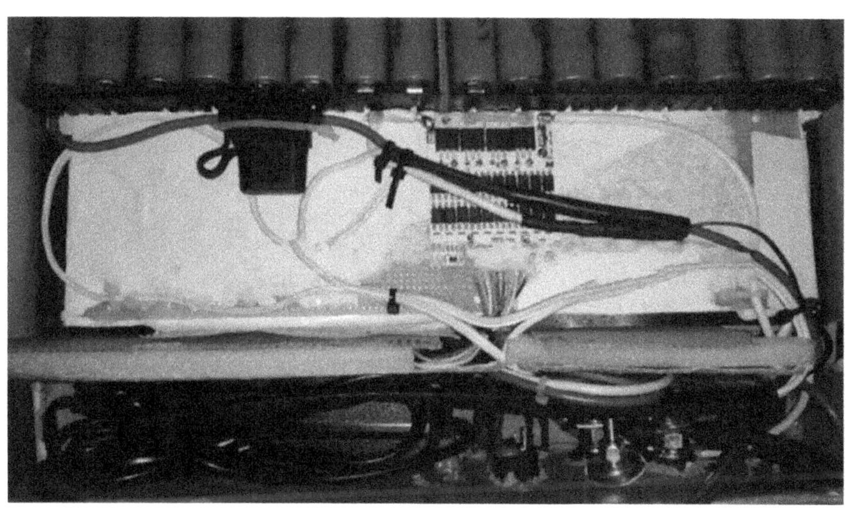

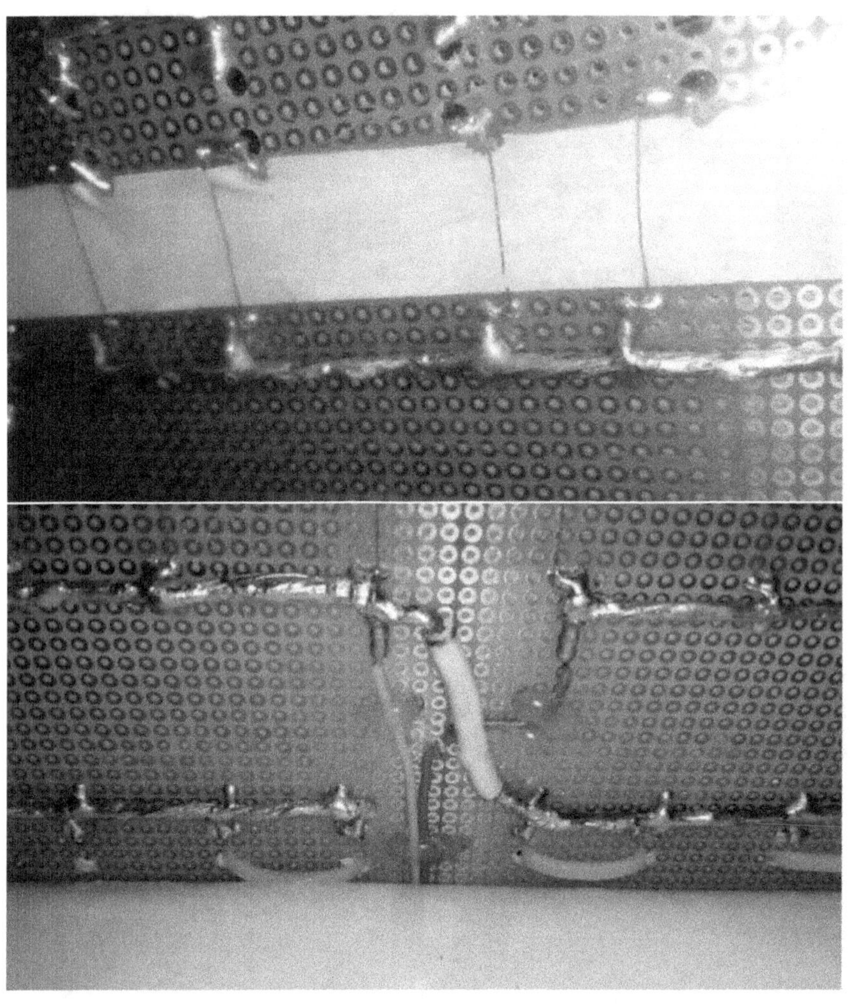

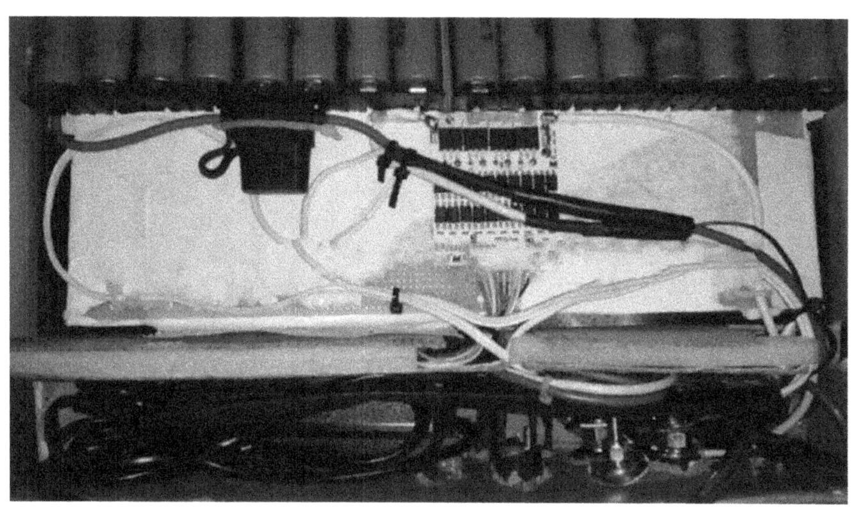

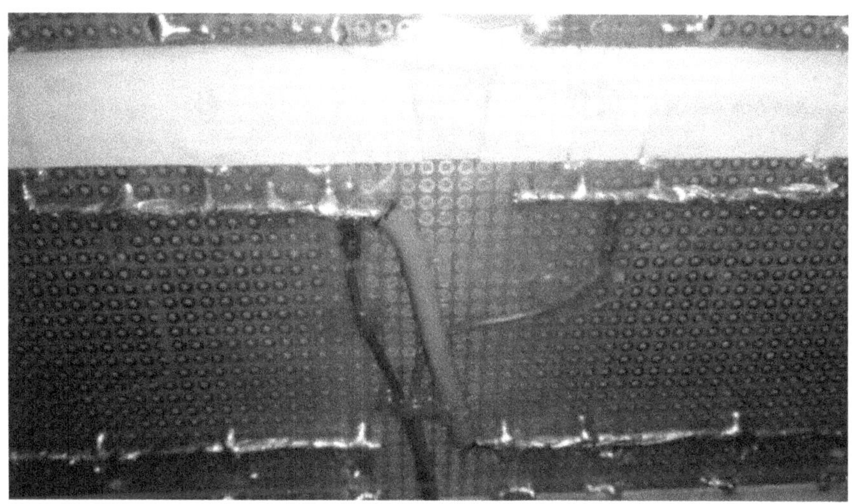

Inside

Divide the inside into two sections: one containing the battery cards and one where you can store everything else.

To prevent the batteries from getting damaged, put some material that absorbs shock while driving.

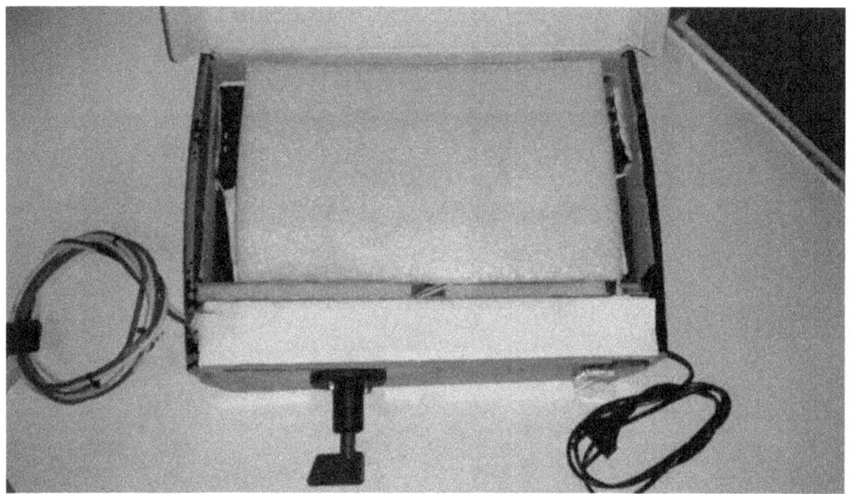

Big Area

Inside the box, in the upper part, you have to place the last two modules of the series of 10 modules together with the BMS, the internal fuse and the related cables following the general scheme.

Divide the box into two areas: a large area where you will house the boards of the 18650 modules, the BMS and the internal fuse and a small area that contains the other components.

Small Area

In the narrowest part of the battery, you can store the charger, the charge indicator display, the battery disconnects switch and the cables connected to them.

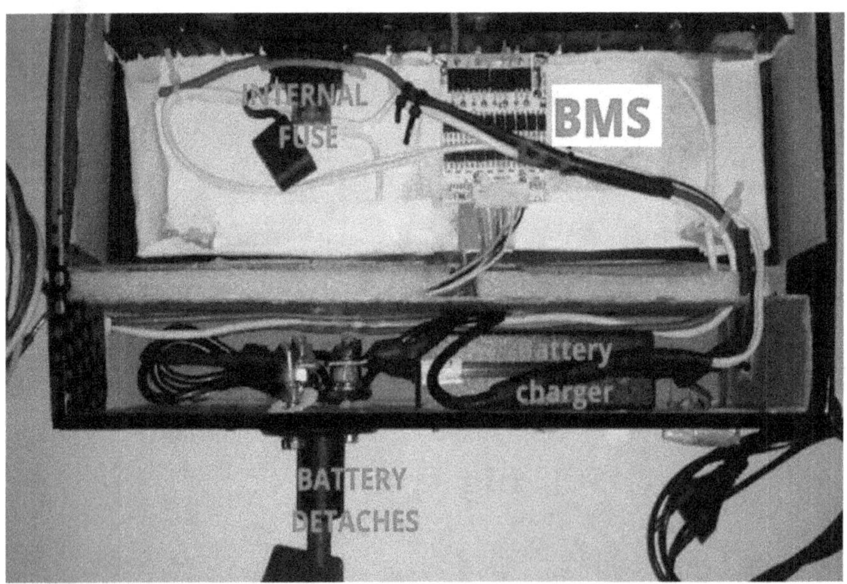

Battery breaker

Cut the positive cable and solder two lugs, one on each side. Then apply the heat shrink sleeve to insulate the whole.

Drill two holes on the front of the box and a rectangular one and insert the base of the battery disconnect switch.

The battery breaker has two poles to which to connect the positive output from the BMS that goes to the engine control unit.

On the left the pole that goes to the engine, on the right the pole that comes from the battery. Tighten the nuts with a suitable key.

Battery Charger

Solder an output plug to the BMS to the battery pack charging cables by soldering the positive to the centre and the negative to the outside.

Bms

The BMS has the following connection points:

- P- Negative pole going to the negative of the engine control unit
- B- To the negative of the whole series of batteries
- CH- Negative coming from the charger
- From B1 to B10 are the positives of each module on the cards.

Internal fuse

The internal fuse has the function of protecting the package from internal faults.

Provide a higher trip amperage of the external fuse so that you do not always have to open the package every time it happens in taste.

You can also put one of 35A.

You have to solder the cables between the positive coming from the battery breaker and the module 10 of the upper board.

Forms

Each card has 4 forms.

Each form includes a group of 8 model 18650 batteries. At the top are all the positives connected in parallel. At the bottom there are all the negatives connected in parallel with each other.

Cards

On each card you have to house 4 battery holder modules, connected in series.

CARD B

Insulation Systems

To isolate the boards from each other, cut rubber sheets of the same size as the boards, 3 mm thick.

For additional insulation and lateral protection of the board size boards leave a margin of 3 cm per side.

It also makes openings to allow the cardboard to bend to the sides.

CHAPTER 13 - Disposal of Old Lithium Batteries

Lithium batteries in smartphones, rechargeable lithium batteries for console controllers and laptop batteries deteriorate over time, decreasing the level of charge you can take advantage of. When the battery no longer holds a charge (it shuts down immediately after charging or goes up to 100% to 10% in a few minutes), we'll have to replace it with a new one to continue using the device. But what to do with the old lithium battery? Where should we throw it away?

Guidelines

Understand the classifications for the disposal of different types of batteries. Batteries contain highly toxic chemicals that are considered hazardous waste. Some of the most common types of batteries and how they are disposed of are:

Alkaline or manganese: This type is used for flashes, toys, remote controls, and smoke alarms. The size ranges from AAA to 9 volts. In the USA, except in California where strict disposal guidelines apply, alkaline batteries are considered common municipal waste and can be disposed of normally.

Zinc Carbon: Considered as sturdy batteries, this type is manufactured in all standard sizes and is not classified as hazardous. Like alkaline batteries, they can be disposed of in the trash.

Button cell: This type of battery is used for hearing aids and watches and contains mercury oxide, lithium, silver oxide or zinc-air. These materials are considered hazardous and should be taken to a domestic hazardous materials collection centre for appropriate treatment.

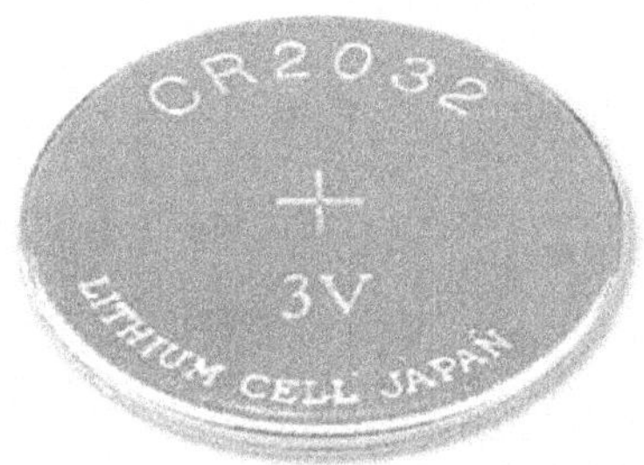

Lithium and lithium-ion: Lithium batteries are used in several small devices and have been labelled as non-hazardous by the government. They are accepted at battery recycling centres.

Rechargeable, alkaline or nickel metal hydride: These types can be disposed of through the normal municipal waste cycle.

Rechargeable, sealed lead-acid or nickel-cadmium: These types must be taken to either a hazardous domestic waste site or a recycling centre.

Lead-acid, vehicle batteries: Car batteries contain sulphuric acid and are 6 or 12 volts. This type is large and contains highly corrosive material. Many vehicle battery retailers will dispose of your old battery when you buy a new one. Metal recyclers will also buy your old battery as waste.

Dispose of your spent batteries properly. The U.S. Environmental Protection Agency and other agencies raise public awareness to bring all discharged batteries to a household hazardous waste collection site or to authorized recycling centres. Batteries that are mistakenly discarded in municipal waste can have serious effects on the environment:

Saturation of landfills, with possible leaching into the soil and infiltration into drinking water tables.

Entry into the atmosphere after incineration. Some metals can be absorbed by the tissues of organisms, with deleterious effects on their health.

Familiarize yourself with the use of environmentally friendly batteries. With a careful and judicious choice, you can choose batteries that have lower levels of heavy metals, reducing the environmental impact in landfills and hazardous waste sites. Some simple steps you can follow are:

Select alkaline batteries when possible. Alkaline battery manufacturers have reduced the amount of mercury since 1984.

Choose silver oxide or zinc-air batteries instead of mercury oxide batteries, which contain higher levels of heavy metals.

Use rechargeable batteries when possible. Recyclable batteries help reduce the environmental impact of dozens of discharged disposable batteries. However, they contain heavy metals.

Buy handheld or solar-powered devices when possible.

CONCLUSION

Thank you for reading this book all the way to the conclusion; we hope it was educational and gave you all the skills you need to reach your objectives, whatever they may be.

This book has attempted to bring all of the crucial aspects to the forefront so that you can obtain all of the knowledge you need on how lithium batteries function as well as how to create a battery pack without having to worry about the bad consequences.

All you have to do is follow the instructions and the information provided in this guide.

By following the easy steps outlined in the book, you can reap all of the benefits of the process.

I hope this guide will really help you achieve your goals.